STEM CELL BIOLOGY IN NORMAL LIFE AND DISEASES

Edited by **Kamran Alimoghaddam**

Stem Cell Biology in Normal Life and Diseases
http://dx.doi.org/10.5772/46173
Edited by Kamran Alimoghaddam

Contributors

Ivan Rich, Atsuko Masumi, Faouzi Jenhani, Yasushi Kubota, Shinya Kimura, Bela Balint, Slobodan Obradovic, Milena Todorovic, Mirjana Pavlovic, Biljana Mihaljevic, Bydlowski, Antonieta Chavez-Gonzalez, Dafne Moreno-Lorenzana, Socrates Aviles-Vazquez, Hector Mayani, Takafumi Yokota, Takao Sudo, Tomohiko Ishibashi, Michiko Ichii, Yukiko Doi, Kenji Oritani, Yuzuru Kanakura, Nasef

Published by InTech

Janeza Trdine 9, 51000 Rijeka, Croatia

Publishing Process Manager Sandra Bakic
Technical Editor InTech DTP team
Cover InTech Design team

Additional hard copies can be obtained from orders@intechopen.com

Stem Cell Biology in Normal Life and Diseases, Edited by Kamran Alimoghaddam
p. cm.
ISBN 978-953-51-1107-8

For EU product safety concerns:
IN TECH d.o.o., Prolaz Marije Krucifikse Kozulić 3, 51000 Rijeka, Croatia,
info@intechopen.com or visit our website at intechopen.com.

Meet the editor

Kamran Alimoghaddam, M.D. was born in 1964 in Tehran (Iran), and his primary research field is in hematology, oncology and bone marrow transplantation. His past positions include: Internship at Hamadan University of Medical sciences, Hamadan (1990-1992); Residency of Internal Medicine at Tehran University of Medical Sciences (1992-1995); Fellowship of Hematology, Oncology and BMT, Tehran University of Medical Sciences (1996-1999); Assistant Professor of Medicine, Tehran University of Medical Sciences (1999-2005); and Associated Professor of Medicine, Tehran University of Medical Sciences (2005-2012). From 2012, Dr. Alimoghaddam is Professor of Medicine, at Tehran University of Medical Sciences. He is a member of EBMT, ESMO, EHA, ASCO, ASH, ECAR, Iranian society of Hematology/Oncology and Iranian society of Bone Marrow Transplantation. Dr. Alimoghaddam is also co-editor of International Journal of Hematology, Oncology and BMT and serves as a member for several editorial boards journals (Hematology, Oncology and Stem Cell Therapy Journal; Khoon (blood) journal; Cell Journal (Yakhteh)).

Contents

Preface

Stem cells have a prominent role in biology of normal life and also in pathogenesis of diseases.

Life starts with generation of an embryo (the most primitive form of a stem cell) and goes through all of the embryonic and fetal stages of life. It forms all of the tissue and organs of a living organism and then has a role in maintenance of normal cellular composition of body. And finally these cells maintain body in a steady state of cell loss and substitution of normal lost/died cells by new normal cells.

Nearly in every organ of body we have these cells, although location characteristic and biological property of them is not completely understood.

Despite of their role in normal life, these cells also have a prominent role in many diseases. Deficiency/depletion of these cells is very important in pathogenesis diseases. For example aplastic anemia (a fatal form of hematopoietic stem cell disease) is stem cell damage in bone marrow, due to environmental or immunologic dysregulation.

Cancer is also a stem cell disease. Genetic/metabolic abnormalities of tissue stem cells (or some times more mature progenitor cells) may disrupt normal proliferation of immature cells in an organ and start process of carcinogenesis.

Although clinical application of the best known stem cells (Hematopoietic stem cells) is possible today and they are useful for treatment of many malignant and non-malignant disorders, clinical knowledge and application of these cells is in infancy. These cells are potentially applicable in regenerative medicine and also cancer treatment.

For safe and effective application of these cells, we need better knowledge of their biology, their interaction with other cells (especially supporting niche cells), growth, maturation and also immigration of stem cells through body in normal and abnormal conditions. Also for clinical application we need to know their characterization, their separation methods and safe manipulation.

This book is written to clarify some aspects of stem cell biology, their characteristics, assessment of damage to cells during ex vivo manipulation and also their role in a model of cancers (chronic myeloid leukemia).

Kamran Alimoghaddam, MD.
Institute for Oncology, Hematology and Stem Cell Transplantation
Tehran University of Medical Sciences, Tehran, Iran

Section 1

Stem Cell Biology

Chapter 1

Regulators of the Proliferation of Hematopoietic Stem and Progenitor Cells During Hematopoietic Regeneration

Yasushi Kubota and Shinya Kimura

Additional information is available at the end of the chapter

http://dx.doi.org/10.5772/55639

1. Introduction

Adult hematopoietic stem cells (HSCs) ensure the maintenance of the HSC pool via their self-renewal capacity, and replenish mature cells throughout life via their proliferation and differentiation into lineage-restricted cells [1, 2]. HSCs are the only stem cells that have been used in the clinic to treat diseases, such as leukemia, germ cell tumors, and congenital immunodeficiencies. Since HSCs were first proposed [3], advances in multicolor flow cytometry have allowed the purification of mouse HSCs close to homogeneity. Several groups have succeeded in long-term hematopoietic reconstitution by transplanting a single lineage of HSCs (e.g., $CD34^{low/-}$ Kit^+ $Sca\text{-}1^+$ lineage marker-negative cells and $CD150^+$ $CD48^-$ Kit^+ $Sca\text{-}1^+$ lineage marker-negative cells), providing direct proof of the existence of HSCs [4-7]. More recently, two groups analyzed the cycling status of HSCs by monitoring their proliferation rate over several months in vivo. The results of these studies suggested the presence of dormant HSCs that divide only about five times throughout the mouse life span [8, 9].

In the bone marrow (BM), HSCs are located in a specialized microenvironment, called the niche. Under steady-state conditions, signals from niches maintain some HSCs in a dormant state. Acquisition of dormancy is critical for the preservation of the self-renewal ability of HSCs and for the prevention of premature stem cell exhaustion [10-12]. However, in response to external stresses, such as bleeding, myeloablative chemotherapy and total body irradiation, HSCs proliferate extensively to produce very large numbers of primitive progenitor cells, thereby enabling rapid hematological regeneration [13]. Once recovery from myelosuppression or other stresses has been achieved, the activated HSCs return to a quiescent state via a number of negative feedback mechanisms [14]. This ability is a hallmark of HSCs and is fundamental for the maintenance of hematopoietic function throughout the life of the organism. Although this property has long been recognized and is very important for organism survival, the molecular basis underlying how HSCs react to a hematologic emergency remains enigmatic. However, some key players have been identified. In this chapter, we briefly review the recent advances in our knowledge of HSC cell-intrinsic and cell-extrinsic regulators that are critical for hematopoietic regeneration under stress hematopoiesis.

2. Cell-intrinsic factors

2.1. Hemeoxygenase-1

Heme plays many important roles, such as in the promotion of proliferation and differentiation of hematopoietic progenitor cells (HPCs) [15], and in the stimulation of hematopoiesis [16, 17]. Heme oxygenase (HO) catalyzes the degradation of heme. HO-1, encoded by the *Hmox1* gene, is a stress-inducible isozyme of HO and is highly expressed in the BM and spleen [17]. Cao et al reported that the hematopoietic lineages of heterozygous HO-1-deficient mice (*HO-1*$^{+/-}$) display accelerated recovery from myelotoxic injury induced by 5-FU treatment. BM transplantation experiments also revealed that mice transplanted with *HO-1*$^{+/-}$ BM cells reconstituted hematopoietic lineages more rapidly than those transplanted with *HO-1*$^{+/+}$ BM cells. However, *HO-1*$^{+/-}$ HSCs could not rescue lethally irradiated recipient mice or serially reconstitute irradiated mice [18]. These results suggest that HO-1 restricts the proliferation and differentiation of HSCs/HPCs under stress conditions, and that the dysregulation of HO-1 can lead to precocious HSC exhaustion.

2.2. PSF-1

PSF-1, a partner of sld five-1, forms a multiprotein complex, termed GINS. The GINS complex contains Psf2, Psf3, and Sld5 in addition to Psf-1 [19-21]. All genes encoding this complex are evolutionally conserved and are essential for cell growth [22]. Ueno et al isolated the mouse orthlog of *PSF1* from a BM Lin$^-$c-Kit$^+$Sca-1$^+$ hematopoietic stem cell cDNA library and found that PSF1 was specifically expressed in immature cells including blastocysts and spermatogonia. They also generated *PSF1*$^{-/-}$ mice lacking functional *PSF1*, which died in utero around the implantation stage [23]. In hematopoietic cells, PSF1 was highly expressed in CD34$^+$ KSL progenitor cells but not in CD34$^-$ KSL cells. In addition, proliferating CD34$^+$ KSL cells sorted from BM 4 days after 5-FU exposure were PSF1 positive. Next, Ueno et al investigated the function of PSF1 in hematopoiesis using *PSF1*$^{+/-}$ mice. The pool size of HSCs and progenitor cells was decreased in aged *PSF1*$^{+/-}$ mice compared to wild-type mice. Whereas young *PSF1*$^{+/-}$ mice showed normal hematopoiesis under steady-state conditions, 5-FU treatment was lethal in *PSF1*$^{+/-}$ mice as a result of a delay in the induction of HSC proliferation after BM ablation [24]. These results suggest that PSF1 is essential for the acute proliferation of HSCs during the regeneration phase after BM suppression.

2.3. Necdin

Necdin belongs to the melanoma antigen family of molecules, whose physiological roles have not been well characterized [25]. Necdin was originally identified as a gene product induced in neurally differentiated embryonal carcinoma cells [26]. Interestingly, recent genetic analyses demonstrate that aberrant genomic imprinting of *NDN* on the human 15q11-q13 chromosomal region is, at least in part, responsible for the pathogenesis of Prader-Willi syndrome [27-29], a disease associated with a mildly increased risk of myeloid leukemia [30]. Necdin interacts with multiple cell cycle-related proteins, such as SV-40 large T antigen, adenovirus E1A, E2F1, and p53 [31-34]. We reported that necdin is one of 32 genes that show higher expression in HSCs than in differentiated hematopoietic cells [35]. Other groups also found that necdin is highly expressed in HSCs [36, 37]. Necdin-deficient mice show accelerated recovery of hematopoietic systems after myelosuppressive stress, such as after 5-FU treatment and BM transplantation, whereas no overt abnormality is seen under conditions of steady-state hematopoiesis. As necdin is a potential negative cell cycle regulator, the enhanced hematologic recovery in necdin-null mice was suggested to result from an increase in the number of proliferating HSCs and progenitor cells. As expected, after 5-FU treatment, necdin-deficient mice had an increased number of HSCs, but this increase was transient and was observed only during the recovery phase [35]. These data suggest that the repression of necdin function in HSCs could form the basis of a novel strategy for the acceleration of hematopoietic recovery, thereby providing therapeutic benefits after clinical myelosuppressive treatments (e.g., cytoablative chemotherapy or HSC transplantation). Necdin is a p53 target gene, and in vitro overexpression and knockdown experiments demonstrated that necdin plays a role in the maintenance of HSC quiescence and self-renewal [37]. However, another group reported that necdin overexpression does not result in enhanced HSC quiescence [38].

2.4. Slug

Slug is a member of the highly conserved Slug/Snail family of zinc-finger transcriptional repressors found in diverse species ranging from C. elegans to humans. SLUG is a target gene for the E2A-HLF chimeric oncoprotein in pro-B cell acute leukemia [39]. Slug is highly expressed in immature hematopoietic cells, and a study using *Slug*$^{-/-}$ mice revealed that slug is essential for the radioprotection of HPCs [40]. Slug is induced by p53 and protects immature hematopoietic cells from apoptosis triggered by DNA damage. Slug exerts this function by repressing Puma, a proapoptotic target of p53 [41]. Sun et al. investigated the effects of Slug under steady-state and stress hematopoiesis [42]. The numbers of HSCs (LSK, Flk2$^-$LSK, SLAM, and EPCR$^+$) and progenitor cells (multipotent progenitors: CD150$^-$CD48$^-$CD244$^+$; lineage-restricted progenitors: CD150$^-$CD48$^+$CD244$^+$) were comparable regardless of the Slug genotype under steady-state conditions. Consistent with previous reports [40, 43], they found that hematopoiesis in the BM was normal in *Slug*$^{-/-}$ mice, suggesting that Slug is not required for steady-state hematopoiesis. On the other hand, an *in vivo* competitive repopulation assay revealed that Slug-deficient BM cells had a higher long-term reconstitution capacity. However, HSC homing and differentiation were not affected by the deficiency of Slug. These results suggest that Slug deficiency increases HSC self-renewal. Next, to assess whether Slug dosage affects HSC self-renewal capability, they performed a serial transplantation assay. *Slug*$^{-/-}$ BM cells showed an enhanced repopulation capacity during serial transplantation. Furthermore, the repopulating and proliferation potential of *Slug*$^{-/-}$ HSCs treated with 5-FU were also examined in vivo and in vitro. Slug deficiency increased the reconstituting potential of 5-FU-activated HSCs in vivo and accelerated HSC expansion in vitro. Taken together, these results suggest that Slug negatively regulates HSC self-renewal under stress hematopoiesis.

2.5. Erg

The E-twenty-six (ETS)-related gene (ERG) belongs to the ETS family of transcription factors [44]. ERG rearrangement has been reported in acute myeloid leukemia (AML) [45] and Ewing sarcoma [46]. ERG overexpression has also been observed in prostate cancer [47]. A recent study revealed that overexpression of ERG is an adverse prognostic factor in AML with a normal karyotype [48]. More recently, the role of Erg in hematopoietic development and normal hematopoiesis was investigated. *Mld2*, an allele of the murine *Erg* gene with a missense mutation in the ETS domain-encoding region, disrupts Erg transactivation of gene expression. Mice homozygous for the *Erg*^{Mld2} allele die at midgestation because of a failure in definitive hematopoiesis [49, 50]. Ng et al studied hematopoiesis in mice heterozygous for the *Erg*^{Mld2} mutation [51]. While *Erg*$^{+/Mld2}$ mice showed normal steady-state hematopoiesis, *Erg*$^{+/Mld2}$ BM cells exhibited defective HSC self-renewal in BM transplantation or during recovery from exposure to sublethal γ-irradiation. The TPO/c-Mpl pathway is critical for the self-renewal and proliferation of HSCs [52-54]. Next, the phenotype of *Erg*$^{+/Mld2}$*Mpl*$^{-/-}$ mice was examined because the Mld2 mutation was originally uncovered during a sensitized ENU mutagenesis screen of Mpl-deficient mice [49]. The double mutant mice died of BM failure following an exacerbation of a defect in HSC proliferation. Thus, Erg is required for HSC self-renewal during stress hematopoiesis. ERG is also expressed in endothelial cells [55, 56]. A recent study showed that ERG plays a role in endothelial tube formation and angiogenesis [57]. More recently, Yuan et al identified Rhoj, a Rho GTPase family member, as a novel downstream target of ERG [58]. Interestingly, Rhoj is also highly expressed in HSCs [35].

3. Cell-extrinsic factors

The interaction between HSCs and their microenvironment (niche) is essential for HSC maintenance, self-renewal, and survival. However, recent studies have revealed that the cell-extrinsic factors provided by the BM niche are also important for HSC responses during hematopoietic regeneration. Some key players have been identified.

3.1. Connexin-43

The exchange of ions, metabolites, and other small molecules (up to ~1,200 Da) occurs via gap junctions. Gap junctions are configured by a large family of proteins known as connexins (Cxs). In the connexin family, connexin-43 (Cx43) is highly expressed in BM stromal cells [59], endothelial cells [7], osteoblasts [60], and mesenchymal stem cells [61]. HSCs also express Cx43

[36]. Cx43-deficiency in the BM (*Mx1*-Cre/*Cx43*$^{flox/flox}$) led to impaired hematopoietic recovery after 5-FU treatment [62]. To clarify the mechanism of impaired hematopoietic regeneration after myeloablation by 5-FU, Taniguchi-Ishikawa et al generated hematopoietic-specific Cx43 (H-Cx43)-deficient mice (*Vav1*-Cre/*Cx43*$^{flox/flox}$) and analyzed their hematopoietic phenotype [63]. The lack of Cx43 in hematopoiesis did not impair long-term competitive repopulation capacity but impaired hematopoietic recovery after 5-FU administration. 5-FU-treated H-Cx43-deficient HSCs failed to enter the cell cycle and showed decreased cell survival. More detailed analyses revealed that enhanced quiescence in H-Cx43-deficient HSCs treated with 5-FU is associated with up-regulation of the expression of quiescence markers, p16^{INK4a} [64] and p38 [65], and with an increased level of intracellular reactive oxygen species (ROS).

The same group also investigated the role of Cx43 in the BM osteoblastic niche. For this, Gonzalez-Nieto et al used conditionally osteoblast lineage-specific Cx43-deficient mice: *Col1-α1*-Cre; *Cx43*$^{flox/flox}$ mice (OB/P Cx43-deficient mice) [66]. The OB/P Cx43-deficient mice showed normal hematopoiesis under steady-state conditions. However, engraftment and migration of normal HSCs was impaired by the loss of Cx43 in the osteoblast lineage. Interestingly, in nonmyeloablated mice, OB/P Cx43 deficiency did not cause a homing defect but increased the endosteal lodgment of HSCs, which was associated with the expansion of Cxcl-12-expressing mesenchymal/osteolineage cells in the BM niche [67]. Another group reported that Cx43 and Cx45 gap junctions mediate the secretion of CXCL12 from BM cells, resulting in HSCs adhesion to stromal cells [68].

3.2. TIMP-3

Metalloproteinases (MMPs) modulate the extracellular matrix (ECM) environment [69-71]. Several studies have indicated that MMP-9 and MT1-MMP are important for the cleavage and inactivation of the KIT ligand (Stem cell factor: SCF) and CXCL12 during G-CSF-induced mobilization and hematopoietic recovery after cytotoxic stress in the BM [72-75]. Tissue inhibitors of the metalloproteinase (TIMP) family consist of four members (TIMP-1 to -4), all of which are endogenous regulators of metalloproteinases (MMPs) [76, 77]. Although TIMPs were initially identified as inhibitors of MMPs, recent findings suggest that they might have more diverse functions [78, 79]. Previous work suggested that TIMP-1 deficiency or enforced expression of TIMP-1 or TIMP-2 does not alter steady-state hematopoiesis and stress hematopoiesis, such as those induced by G-CSF stimulation and myelotoxic insult, respectively [80]. However, Rossi et al recently found that increased expression of p53 in *TIMP-1*$^{-/-}$ HSCs resulted in dysregulation of the transition from G1 to S phase of the cell cycle, indicating that TIMP-1 has a role in controlling the cell cycle dynamics of LT-HSCs [81].

Among TIMP family members, TIMP-3 has unique properties. TIMP-3 binds firmly to ECM, a disintegrin [82-84], and inhibits metalloproteinase domain-containing proteins, such as ADAMs and ADAMTSs (ADAM proteins with Thrombospondin Motifs) [85]. Nakajima et al found decreased expression of TIMP-3 in immune suppressor factor (ISF)/short form of ISF (ShIF)-transfected cell lines, and partial reduction of HSC-supporting activity following the restoration of TIMP-3 expression in stromal cells expressing ISF [86]. These authors also investigated the role of TIMP-3 in HSC regulation. TIMP-3 expression in BM was increased after 5-FU injection, and addition or overexpression of TIMP-3 resulted in enhanced proliferation of HSCs in vitro. BM regeneration after myelotoxic stress was impaired in TIMP-3-deficient mice, but was accelerated by enforced expression of TIMP-3 in vivo [87]. Another group also studied the role of TIMP-3 in hematopoiesis. They found that TIMP-3 was highly expressed in the endosteal region of the BM, the HSC niche, whereas its expression was low in HSCs and progenitor cells. They also examined the effect of human TIMP-3 (huTIMP-3) overexpression in HSCs in vivo. TIMP-3 overexpression resulted in increased myelopoiesis and decreased lymphopoiesis. Consistent with the study of Nakajima et al [87], HSC proliferation was increased by huTIMP-3 overexpression in vitro and in vivo [88]. These results suggest that TIMP-3 is important for the cellular response to myelosuppression.

3.3. Tenascin-C

Tenascin-C (TN-C) is a large extracellular matrix (ECM) glycoprotein that is expressed mainly in the developing embryo [89]. In the adult BM, expression of TN-C is restricted to the endosteal region [90, 91]. Although TN-C-deficient mice exhibit grossly normal development, the colony-forming capacity of $TN\text{-}C^{-/-}$ BM cells is lower than that of wild-type BM cells [92]. This suggests that TN-C makes a significant contribution under stress hematopoiesis because the mononuclear cell count and BM architecture of TN-C-deficient mice are essentially normal. Nakamura-Ishizu et al studied the function of TN-C during hematopoiesis in vivo using TN-C knockout mice. First, they examined the expression pattern of various ECM proteins in the BM under different conditions (steady-state, immediately after myeloablation, and during the hematopoietic recovery phase). TN-C was predominantly expressed in stromal cells and endothelial cells, which are components of the BM niche, and was markedly up-regulated in the BM during hematopoietic regeneration. $TN\text{-}C^{-/-}$ mice showed defects in hematopoietic recovery after BM ablation caused by 5-FU treatment and sublethal irradiation. The transplantation of wild-type BM cells into $TN\text{-}C^{-/-}$ recipient mice demonstrated that a supporting ability of hematopoiesis in BM microenvironment lacking TN-C is inadequate for the proliferation of transplanted wild-type BM cells for the regeneration of hematopoiesis [93]. These findings suggest that TN-C is a critical component of the BM microenvironment during hematopoietic regeneration.

4. Conclusions

In this chapter, we have briefly summarized regulators that have recently been shown to be involved in the control of HSCs and progenitors during the hematopoietic regeneration phase. Recent studies have revealed that, when myelopoiesis is compromised following infection, HSC proliferation involves not only the factors described in this review, but also inflammatory signaling molecules such as interferons [94-96], tumor necrosis factor-α [97, 98], and Toll-like receptors [99-101]. Because HSC proliferation potential is critical for organism survival during stress conditions, further understanding of the mechanism of stem cell activation will be needed before stem cells can be used in regenerative medicine.

http://dx.doi.org/10.5772/55639

Acknowledgements

This work was supported by a Grant-in-Aid for Young Scientists to Y.K. (no. 23791083) from the Ministry of Education, Culture, Sports, Science and Technology (MEXT), Japan.

Author details

Yasushi Kubota[1,2*] and Shinya Kimura[1]

*Address all correspondence to: kubotay@cc.saga-u.ac.jp

1 Division of Hematology, Respiratory Medicine and Oncology, Department of Internal Medicine, Faculty of Medicine, Saga University, Japan

2 Department of Transfusion Medicine, Saga University Hospital, Japan

References

[1] Orkin, S.H., & Zon, L. I. Hematopoiesis: An Evolving Paradigm for Stem Cell Biology. Cell 2008; 132(4): 631-644.

[2] Kondo, M., Wagers, A.J., Manz, M.G., Prohaska, S.S., Scherer, D.C., Beilhack, G.F., et al. Biology of hematopoietic stem cells and progenitors: implications for clinical application. Annu Rev Immunol 2003; 21: 759-806.

[3] Till, J.E., & McCulloch, C. E. A direct measurement of the radiation sensitivity of normal mouse bone marrow cells. Radiat Res 1961; 14(2): 213-222.

[4] Osawa, M., Hanada, K., Hamada, H. & Nakauchi, H. Long-term lymphohematopoietic reconstitution by a single CD34$^{-low/negative}$ hematopoietic stem cell. Science 1996; 273(5272): 242-245.

[5] Wagers, A. J., Sherwood, R. I., Christensen, J. L. & Weissman, I. L. Little evidence for developmental plasticity of adult hematopoietic stem cells. Science 2002; 297(5590): 2256-2259.

[6] Matsuzaki, Y., Kinjo, K., Mulligan, R.C. & Okano, H. Unexpectedly efficient homing capacity of purified murine hematopoietic stem cells. Immunity 2004; 20(1): 87-93.

[7] Kiel, M.J., Yilmaz, O.H., Iwashita, T., Terhorst, C. & Morrison, S.J. SLAM family receptors distinguish hematopoietic stem and progenitor cells and reveal endosteal niches for stem cells. Cell 2005; 121(7): 1109-1121.

[8] Wilson, A., Laurenti, E., Oser, G., et al. Hematopoietic stem cells reversibly switch from dormancy to self-renewal during homeostasis and repair. Cell 2008; 135(6): 1118-1129.

[9] Foudi, A., Hochedlinger, K., Van Buren, D., et al. Analysis of histone 2B-GFP retention reveals slowly cycling hematopoietic stem cells. Nat Biotechnol 2009; 27(1): 84-90.

[10] Wilson, A., & Trumpp, A. Bone-marrow haematopoietic-stem-cell niches. Nat Rev Immunol 2006 ;6(2): 93-106.

[11] Kiel, M.J., & Morrison, S.J. Uncertainty in the niches that maintain haematopoietic stem cells. Nat Rev Immunol 2008; 8(4): 290-301.

[12] Arai, F., Yoshihara, H., & Hosokawa, K., et al. Niche regulation of hematopoietic stem cells in the endosteum. Ann N Y Acad Sci 2009;1176: 36-46.

[13] Randall, T.D., & Weissman, I.L. Phenotypic and functional changes induced at the clonal level in hematopoietic stem cells after 5-fluorouracil treatment. Blood 1997; 89(10): 3596-3606.

[14] Venezia, T.A., Merchant, A.A., Ramos, C.A., et al. Molecular signatures of proliferation and quiescence in hematopoietic stem cells. PLoS Biol 2004; 2(10): e301.

[15] Chertkov, J.L., Jiang, S., Lutton, J.D., Levere, R.D. & Abraham, N.G. Hemin stimulation of hematopoiesis in murine long-term bone marrow culture. Exp Hematol 1991; 19(9): 905-909.

[16] Porter, P.N., Meints, R.H., & Mesner, K. Enhancement of erythroid colony growth in culture by hemin. Exp Hematol 1979; 7(1): 11-16.

[17] Abraham, N.G. Molecular regulation—biological role of heme in hematopoiesis. Blood Rev 1991; 5(1): 19-28.

[18] Cao, Y.A., Wagers, A.J., Karsunky, H., et al. Heme oxygenase-1 deficiency leads to disrupted response to acute stress in stem cells and progenitors. Blood 2008; 112(12): 4494-4502.

[19] Kanemaki, M., Sanchez-Diaz, A., Gambus, A. & Labib, K. Functional proteomic identification of DNA replication proteins by induced proteolysis in vivo. Nature 2003; 423(6941): 720-724.

[20] Kubota, Y., Takase, Y., Komori, Y., et al. A novel ring-like complex of Xenopus proteins essential for the initiation of DNA replication. Genes Dev 2003; 17(9): 1141-1152.

[21] Takayama, Y., Kamimura, Y., Okawa, M., Muramatsu, S., Sugino, A. & Araki, H. GINS, a novel multiprotein complex required for chromosomal DNA replication in budding yeast. Genes Dev 2003; 17(9): 1153-1165.

[22] Winzeler, E. A., Shoemaker, D.D., Astromoff, A., et al. Functional characterization of the S. cerevisiae genome by gene deletion and parallel analysis. Science 1999: 285(5429): 901-906.

[23] Ueno, M., Itoh, M., Kong, L., Sugihara, K., Asano, M. & Takakura, N. PSF1 is essential for early embryogenesis in mice. Mol Cell Biol 2005; 25(23): 10528-10532.

[24] Ueno, M., Itoh, M., Sugihara, K., Asano, M., & Takakura, N. Both alleles of PSF1 are required for maintenance of pool size of immature hematopoietic cells and acute bone marrow regeneration. Blood 2009; 113(3): 555-562.

[25] Xiao, J., & Chen, H.S. Biological functions of melanoma-associated antigens. World J Gastroenterol 2004; 10(13): 1849-1853.

[26] Maruyama, K., Usami, M., Aizawa, T., & Yoshikawa, K. A novel brain-specific mRNA encoding nuclear protein (necdin) expressed in neurally differentiated embryonal carcinoma cells. Biochem Biophys Res Commun 1991; 178(1): 291-296.

[27] MacDonald, H.R., & Wevrick, R. The necdin gene is deleted in Prader-Willi syndrome and is imprinted in human and mouse. Hum Mol Genet 1997; 6(11): 1873-1878.

[28] Nakada, Y., Taniura, H., Uetsuki, T., Inazawa, J., & Yoshikawa, K. The human chromosomal gene for necdin, a neuronal growth suppressor, in the Prader-Willi syndrome deletion region. Gene 1998; 213(1-2): 65-72.

[29] Barker, P.A., & Salehi, A. The MAGE proteins: emerging roles in cell cycle progression, apoptosis, and neurogenetic disease. J Neurosci Res 2002; 67(6): 705-712.

[30] Davies, H.D., Leusink, G.L., McConnell, A., et al. Myeloid leukemia in Prader-Willi syndrome. J Pediatr 2003; 142(2): 174-178.

[31] Taniura, H., Taniguchi, N., Hara, M., & Yoshikawa, K. Necdin, a postmitotic neuron-specific growth suppressor, interacts with viral transforming proteins and cellular transcription factor E2F1. J Biol Chem 1998; 273(2): 720-728.

[32] Taniura, H., Matsumoto, K., & Yoshikawa, K. Physical and functional interactions of neuronal growth suppressor necdin with p53. J Biol Chem 1999; 274(23): 16242-16248.

[33] Taniura, H., Kobayashi, M., & Yoshikawa, K. Functional domains of necdin for protein-protein interaction, nuclear matrix targeting, and cell growth suppression. J Cell Biochem 2005; 94(4): 840-815.

[34] Hu, B., Wang, S., Zhang, Y., Feghali, C.A., Dingman, J.R., & Wright, T.M. A nuclear target for interleukin-1alpha: interaction with the growth suppressor necdin modulates proliferation and collagen expression. Proc Natl Acad Sci U S A 2003; 100(17): 10008-10013.

[35] Kubota, Y., Osawa, M., Jakt, L.M., Yoshikawa, K., & Nishikawa, S-I. Necdin restricts proliferation of hematopoietic stem cells during hematopoietic regeneration. Blood 2009; 114(20): 4383-4392.

[36] Forsberg, E.C., Prohaska, S.S., Katzman, S., Heffner, G.C., Stuart, J.M., & Weissman, I.L. Differential expression of novel potential regulators in hematopoietic stem cells. PLoS Genet 2005; 1(3): e28.

[37] Liu, Y., Elf, S.E., Miyata, Y., et al. p53 regulates hematopoietic stem cell quiescence. Cell Stem Cell 2009; 4(1): 37-48.

[38] Sirin, O., Lukov, G.L., Mao, R., Conneely, O.M., & Goodell, M.A. The orphan nuclear receptor Nurr1 restricts the proliferation of haematopoietic stem cells. Nat Cell Biol 2010; 12(12): 1213-1219.

[39] Inukai, T., Inoue, A., Kurosawa, H., et al. SLUG, a ces-1-related zinc finger transcription factor gene with antiapoptotic activity, is a downstream target of the E2A-HLF oncoprotein. Mol Cell 1999; 4(3): 343-352.

[40] Inoue, A., Seidel, M.G., Wu, W., et al. Slug, a highly conserved zinc finger transcriptional repressor, protects hematopoietic progenitor cells from radiation-induced apoptosis in vivo. Cencer cell 2002; 2(4): 279-288.

[41] Wu, W.S., Heinrichs, S., Xu, D., et al. Slug antagonizes p53-mediated apoptosis of hematopoietic progenitors by repressing puma. Cell 2005; 23(4): 641-653.

[42] Sun, Y., Shao, L., Bai, H., Wang, Z.Z. & Wu, W.S. Slug deficiency enhances self-renewal of hematopoietic stem cells during hematopoietic regeneration. Blood 2010; 115(9): 1709-1717.

[43] Perez-Losada, J., Sanchez-Martin, M., Rodriguez-Garcia, et al. Zinc-finger transcription factor Slug contributes to the function of the stem cell factor c-kit signaling pathway. Blood 2002; 100(4): 1274-1286.

[44] Reddy, E.S., Rao, V.N. & Papas, T.S. The erg gene: a human gene related to the ets oncogene. Proc Natl Acad Sci U S A 1987; 84(17): 6131-6135.

[45] Ichikawa, H., Shimizu, K., Hayashi, Y. & Ohki, M. An RNA-binding protein gene, TLS/FUS, is fused to ERG in human myeloid leukemia with t(16;21) chromosomal translocation. Cancer Res 1994; 54(11): 2865-2868.

[46] Sorensen, P.H.B., Lessnick, S.L., Lopez-Terrada, D., Liu, X.F., Triche, T.J. & Denny, C.T. A second Ewing's sarcoma translocation, t(21;22), fuses the EWS gene to another ETS-family transcription factor, ERG. Nat Genet 1994; 6(2): 146-151.

[47] Petrovics, G., Liu, A., Shaheduzzaman, S., et al. Frequent overexpression of ETS-related gene-1 (ERG1) in prostate cancer trascriptome. Oncogene 2005; 24(23): 3847-3852.

[48] Marcucci, G., Baldus, C.D., Ruppert, A.S., et al. Overexpression of the ETS-related gene, ERG, predicts a worse outcome in acute myeloid leukemia with normal karyotype: a Cancer and Leukemia Group B study. J Clin Oncol 2005; 23(36): 9234-9242.

[49] Loughran, S.J., Kruse, E.A., Hacking, D.F., et al. The transcription factor Erg is essential for definitive hematopoiesis and the function of adult hematopoietic stem cells. Nat Immunol 2008; 9(7): 810-819.

[50] Taoudi, S., Bee, T., Hilton, A., et al. ERG dependence distinguishes developmental control of hematopoietic stem cell maintenance from hematopoietic specification. Genes Dev 2011; 25(3): 251-262.

[51] Ng, A.P., Loughran, S.J., Metcalf, D., et al. Erg is required for self-renewal of hematopoietic stem cells during stress hematopoiesis in mice. Blood 2011; 118(9): 2454-2461.

[52] Kimura, S., Roberts, A.W., Metcalf, D., & Alexander, W.S. Hematopoietic stem cell deficiencies in mice lacking c-Mpl, the receptor for thrombopoietin. Proc Natl Acad Sci USA 1998; 95(3): 1195-1200.

[53] Qian, H., Buza-Vidas, N., Hyland, C.D., et al. Critical role of thrombopoietin in maintaining adult quiescent hematopoietic stem cells. Cell Stem Cell 2007; 1(6): 671-684.

[54] Yoshihara, H., Arai, F., Hosokawa, K., et al. Thrombopoietin/MPL signaling regulates hematopoietic stem cell quiescence and interaction with the osteoblastic niches. Cell Stem Cell 2007; 1(6): 685-697.

[55] Baltzinger, M., Mager-Heckel, A.M., & Remy, P, Xl. erg: expression pattern and overexpression during development plead for a role in endothelial cell differentiation. Dev Dyn 1999; 216(4–5): 420-433.

[56] Hewett, P.W., Nishi, K., Daft, E.L., & Clifford Murray, J. Selective expression of erg isoforms in human endothelial cells. Int J Biochem Cell Biol 2001; 33(4): 347-355.

[57] Birdsey, G.M., Dryden, N.H., & Amsellem, V., et al. Transcription factor Erg regulates angiogenesis and endothelial apoptosis through VE-cadherin. Blood 2008; 111(7): 3498-3506.

[58] Yuan, L., Sachiaridou, A., Stratman, A.N., et al. RhoJ is an endothelial cell-restricted Rho GTPase that mediates vascular morphogenesis and is regulated by the transcription factor ERG. Blood 2011; 118(4): 1145-1153

[59] Cancelas, J.A., Koevoet, W.L., de Koning, A.E., Mayen, A.E., Rombouts, E.J., & Ploemacher, R.E. Connexin-43 gap junctions are involved in multiconnexin-expressing stromal support of hemopoietic progenitors and stem cells. Blood. 2000; 96(2): 498-505.

[60] Civitelli, R. Cell-cell communication in the osteoblast/osteocyte lineage. Arch Biochem Biophys 2008; 473(2): 188-192.

[61] Mendez-Ferrer, S., Michurina, T.V., Ferraro, F., et al. Mesenchymal and haematopoietic stem cells form a unique bone marrow niche. Nature 2010; 466(7308): 829-834.

[62] Presley, C,A., Lee, A.W., Kastl, B., et al. Bone marrow connexin-43 expression is critical for hematopoietic regeneration after chemotherapy. Cell Commun Adhes 2005; 12(5-6): 307-317.

[63] Taniguchi-Ishikawa, E., Gonzalez-Nieto, D., Ghiaur, G., et al. Connexin-43 prevents hematopoietic stem cell senescence through transfer of reactive oxygen species to bone marrow stromal cells. Proc Natl Acad Sci U S A 2012; 109(23): 9071-9076.

[64] Janzen, V., Forkert, R., Fleming, H.E., et al. Stem-cell ageing modified by the cyclin-dependent kinase inhibitor p16INK4a. Nature 2006; 443(7110): 421-426.

[65] Ito, K., Hirao, A., Arai, F., et al. Reactive oxygen species act through p38 MAPK to limit the lifespan of hematopoietic stem cells. Nat Med 2006; 12(4): 446-451.

[66] Chung, D.J., Castro, C.H., Watkins, M., et al. Low peak bone mass and attenuated anabolic response to parathyroid hormone in mice with an osteoblast-specific deletion of connexin43. J Cell Sci 2006; 119(Pt20): 4187-4198.

[67] Gonzalez-Nieto, D., Li, L., Kohler, A., et al. Connexin-43 in the osteogenic BM niche regulates its cellular composition and the bidirectional traffic of hematopoietic stem cells and progenitors. Blood 2012; 119(22): 5144-5154.

[68] Schajnovitz, A., Itkin, T., D'Uva, G., et al. CXCL12 secretion by bone marrow stromal cells is dependent on cell contact and mediated by connexin-43 and connexin-45 gap junctions. Nat Immunol 2011; 12(5): 391-398.

[69] Cruz, A.C., Frank, B.T., Edwards, S.T., Dazin, P.F., Peschon, J.J., & Fang, K.C. Tumor necrosis factor-alpha-converting enzyme controls surface expression of c-Kit and survival of embryonic stem cell-derived mast cells. J Biol Chem 2004; 279(7): 5612-5620.

[70] Mezyk, R., Bzowska, M., & Bereta, J. Structure and functions of tumor necrosis factor-alpha converting enzyme. Acta Biochim Pol 2003; 50(3): 625-645.

[71] Baumann, G., Frank, S.J. Metalloproteinases and the modulation of GH signaling. J Endocrinol 2002; 174(3): 361-368.

[72] Petit, I., Szyper-Kravitz, M., Nagler, A., et al. G-CSF induces stem cell mobilization by decreasing bone marrow SDF-1 and up-regulating CXCR4. Nat Immunol 2002; 3(7): 687–694.

[73] Pruijt, J.F., Fibbe, W.E., Laterveer, L., et al. Prevention of interleukin-8-induced mobilization of hematopoietic progenitor cells in rhesus monkeys by inhibitory antibodies against the metalloproteinase gelatinase B (MMP-9). Proc Natl Acad Sci USA 1999; 96(19): 10863–10868.

[74] McQuibban, G.A., Butler, G.S., Gong, J.H., et al. Matrix metalloproteinase activity inactivates the CXC chemokine stromal cell-derived factor-1. J Biol Chem 2001; 276(47): 43503–43508.

[75] Heissig, B., Hattori, K., Dias, S., et al. Recruitment of stem and progenitor cells from the bone marrow niche requires MMP-9 mediated release of kit-ligand. Cell 2002; 109(5): 625–637.

[76] Gomez, D.E., Alonso, D.F., Yoshiji, H. & Thorgeisson, U.P. Tissue inhibotors of metalloproteinases: structure, regulation and biological functions. Eur J Cell Biol 1997; 74(2): 111-122.

[77] Nagase, H., Visse, R., Murphy, G. Structure and function of matrix metalloproteinases and TIMPs. Cardiovasc Res 2006; 69(3): 562-573.

[78] Lambert, E., Dassé, E., Haye, B., & Petitfrère, E. TIMPs as multifacial proteins. Crit Rev Oncol Hematol 2004; 49(3): 187-198.

[79] Chirco, R., Liu, X.W., Jung, K.K., & Kim, H.R. Novel functions of TIMPs in cell signaling. Cancer Metastasis Rev 2006; 25(1): 99-113.

[80] Haviernik, P., Diaz, M.T., Haviernikova, E., Tse, W., Stetler-Stevenson, W.G., & Bunting, K.D. Hematopoiesis in mice is extremely resilient to wide variation in TIMP/MMP balance. Blood Cells Mol Dis 2008; 41(2): 179-187.

[81] Rossi, L., Ergen, A.V., & Goodell, M.A. TIMP-1 deficiency subverts cell-cycle dynamics in murine long-term HSCs. Blood 2011; 117(24): 6479-6488.

[82] Pavloff, N., Staskus, P.W., Kishnani, N.S., & Hawkes, S.P. A new inhibitor of metalloproteinases from chicken: ChIMP-3. A third member of the TIMP family. J Biol Chem 1992; 267(24): 17321–17326.

[83] Yu, W.H., Yu, S., Meng, Q., Brew, K., & Woessner, J.F.Jr. TIMP-3 binds to sulfated glycosaminoglycans of the extracellular matrix. J Biol Chem 2000; 275(40): 31226–31232.

[84] Lee, M.H., Atkinson, S., & Murphy, G. Identification of the extracellular matrix (ECM) binding motifs of tissue inhibitor of metalloproteinases (TIMP)-3 and effective transfer to TIMP-1. J Biol Chem 2007; 282(9): 6887–6898.

[85] Woessner, J.F. Jr. That impish TIMP: the tissue inhibitor of metalloproteinases-3. J Clin Invest 2001; 108(6): 799-800.

[86] Nakajima, H., Shibata, F., & Fukuchi, Y., et al. Immune suppressor factor confers stromal cell line with enhanced supporting activity for hematopoietic stem cells. Biochem Biophys Res Commun 2006; 340(1): 35-42.

[87] Nakajima, H., Ito, M., & Smookler, D.S., et al. TIMP-3 recruits quiescent hematopoietic stem cells into active cell cycle and expands multipotent progenitor pool. Blood 2010; 116(22): 4474-4482.

[88] Shen, Y., Winkler, I.G., Barbier, V., Sims, N.A., Hendy, J., & Lévesque, J.P. Tissue inhibitor of metalloproteinase-3 (TIMP-3) regulates hematopoiesis and bone formation in vivo. PLoS One. 2010; 5(9). doi:pii: e13086.

[89] Hsia, H.C., & Schwarzbauer, J.E. Meet the tenascins: multifunctional and mysterious. J Biol Chem 2005; 280(29): 26641-26644.

[90] Klein, G., Beck, S., & Muller, C.A. Tenescin is a cytoadhesive extracellular matrix component of the human hematopoietic microenvironment. J Cell Biol 1993; 123(4): 1027-1035.

[91] Soini, Y., Kamel, D., Apaja-Sarkkinen, M., Virtanen, I., & Lehto, V.P. Tenascin immunoreactivity in normal and pathological bone marrow. J Clin Pathol 1993; 46(3): 218-221.

[92] Ohta, M., Sakai, T., Saga, Y., Aizawa, S-I., Saito, M. Suppression of hematopoietic activity in Tenascin-C-deficient mice. Blood 1998; 91(11): 4074-4083.

[93] Nakamura-Ishizu, A., Okuno, Y., Omatsu, Y., et al. Extracellular matrix protein tenascin-C is required in the bone marrow microenvironment primed for hematopoietic regeneration. Blood 2012; 119(23): 5429-5437.

[94] Essers, M.A., Offner, S., Blanco-Bose, W.E., et al. IFNalpha activates dormant haematopoietic stem cells in vivo. Nature 2009; 458(7240): 904-908.

[95] Zhao, X., Ren, G., Liang, L., et al. Brief report: interferon-gamma induces expansion of Lin(-)Sca-1(+)C-Kit(+) Cells. Stem Cells 2010; 28(1): 122-126.

[96] Baldridge, M.T., King, K.Y., Boles, N.C., Weksberg, D.C., & Goodell, M.A. Quiescent haematopoietic stem cells are activated by IFN-gamma in response to chronic infection. Nature 2010; 465(7299): 793-797.

[97] Rezzoug, F., Huang, Y., Tanner, M.K., et al. TNF-α is critical to facilitate hematopoietic stem cell engraftment and function. J Immunol 2008; 180(1): 49-57.

[98] Rebel, V.I., Hartnett, S., Hill, G.R., Lazo-Kallanian, S.B., Ferrara, J.L., & Sieff, C.A. Essential role for the p55 tumor necrosis factor receptor in regulating hematopoiesis at a stem cell level. J Exp Med 1999; 190(10): 1493-1504.

[99] Nagai, Y., Garrett, K.P., Ohta, S., et al. Toll-like receotors on hematopoietic progenitor cells stimulate innate immune system replenishment. Immunity 2006; 24(6): 801-812.

[100] Sioud, M., Floisand, Y., Forfang, L, & Lund-Johansen, F. Signaling through toll-like receptor 7/8 induces the differentiation of human bone marrow CD34+ progenitor cells along the myeloid lineage. J Mol Biol 2006; 364(5): 945-954.

[101] Sioud, M., & Floisand, Y. TLR agonists induce the differentiation of human bone marrow CD34+ progenitor cells into CD11c+ CD80/86+ DC capable of inducing a Th1-type response. Eur J Immunol 2007; 37(10): 2834-2846.

Chapter 2

Hematopoietic Stem Cell Niche: Role in Normal and Malignant Hematopoiesis

Sergio P. Bydlowski, Debora Levy,
Jorge M.L. Ruiz and Juliana Pereira

Additional information is available at the end of the chapter

http://dx.doi.org/10.5772/55508

1. Introduction

A stem-cell niche can be defined as a spatial structure in which stem cells are housed and maintained by allowing self-renewal in the absence of differentiation. The concept of the stem-cell niche was initially proposed in invertebrates where they were first characterized; the control of germ line stem cell maintenance and differentiation was observed in ovarian and testicular *D. melanogaster* niches (Wilson & Trumpp, 2006).

Hematological stem cells (HSCs) can be initially identified in the developing embryo in aorta-gonada-mesonephro regions, form where they migrate into the fetal liver and subsequently move into the bone marrow (BM) after birth. During development, HSCs undergo active self-renewal to expand the pool size, but then become largely quiescent and stay in a steady state in adult BM, where their maintenance is tightly regulated (Oh & Kwon, 2010).

The link between bone-marrow formation (hematopoiesis) and bone development (osteogenesis) was first recognized in the 1970s. It was noted that whilst HSCs in the bone marrow drive hematopoiesis during the whole life of organisms, when they are removed from the bone marrow, they lose the ability to self-renewal, indicating similar dependence of HSCs on extrinsic signals (Lilly *et al.*, 2011). The term 'niche' used for the specific HSC bone-marrow microenvironment was first coined by Schofield (1978), who observed that HSC growth was not supported in the spleen in the same manner as in the bone marrow; he also proposed that HSCs are in intimate contact with bone, and that cell–cell contact was responsible for the apparently unlimited proliferative capacity and inhibition of maturation of HSCs.

This concept has since been further developed, and it is now widely accepted that specific anatomical regions within the bone marrow comprise specialized niches for HSC development and normal blood cell production (Noll *et al.*, 2012). The niche is composed of a vast milieu of cellular and humoral factors.

The advancement of our understanding of the interactions between bone, the microenvironment and hematopoiesis is rapidly accelerating. Research continues on the identification and characterization of individual cell populations constituting the hematopoietic stem cell (HSC) niche. Whether there are one or multiple HSC niches and how cells within the HSC niche interact with each other are also subjects of interest (Shen *et al.*, 2012).

Recently it has been suggested that two main microenvironments form the bone marrow niches: 1. the endosteal niche, where osteoblasts derived from mesenchymal precursors are localized in the endosteal regions, and 2. The vascular niche, which might be involved in HSC maintenance within the bone marrow (Chotinantakul *et al.*, 2012).

2. The endosteal niche

The first direct evidence for cells having stem cell-supporting activity in bone formation was provided by studies in which both mouse and human osteoblast cell lines were shown to secrete a large number of cytokines that promote the proliferation of hematopoietic cells in culture (Wilson & Trumpp, 2006). Studies in the 1970s indicated that undifferentiated hematopoietic cells are localized close to the endosteal bone surface, and that differentiated cells move toward the central axis of the marrow. The endosteum is the interface between bone and bone marrow and constitutes a reservoir of HSCs that can be mobilized, restoring hematopoiesis in response to tissue injury. Although bone is a hard and stiff organ, the endosteal niche is not so; it rather presents plasticity and is under self and systemic regulation.

After the identification of niches in the endosteal region, cells in the endothelial lining were proposed as a cellular component of the niche. However, heterogeneous cell populations were found in endosteal regions that included mature bone lining cells, osteoblasts, and preosteoblasts (Oh *et al.*, 2010).

2.1. Osteoblastic cells

Osteoblasts are responsible for the production of matrix, which they secrete at the site of the bone, as well as for bone mineralization. Eventually, osteoblasts either get surrounded by matrix and end up as osteocytes or they become bone-lining cells, which is a reversible process (ter Ruurne *et al.*, 2010). The bone-forming osteoblastic cells are crucial players for the homeostasis of the hematopoietic tissue with its high turnover. In 1994, primary human osteoblasts were shown to stimulate the proliferation of primitive CD34+ hematopoietic progenitors in vitro, which raised the possibility that osteoblasts are involved in hematopoiesis (Nagasawa *et al.*, 2011).

Involvement of osteoblastic cells in HSC regulation and maintenance in vivo was first reported by two groups using engineered mouse models (Lymperi *et al.*, 2010). The first study, a bone morphogenic protein (BMP) receptor IA (BMPRIA) conditional knockout mouse was used to show that an increase in the number of osteoblastic cells was correlated with an increase in the number of HSCs.

HSCs proliferation and differentiation is regulated by a wide variety of cytokines, growth factors, and other signaling molecules. Osteoblastic cells synthesize a number of cell-signaling molecules that appear to contribute to the maintenance and regulation of HSCs by the endosteal niche such as (table1):

- Jagged, a ligand for Notch receptors expressed on HSCs which is markedly upregulated with activation of the osteoblast by the parathyroid hormone. Activation of Notch receptors on HSCs has been shown to inhibit differentiation and enhance their self renewal capacity in vitro (Butler *et al.,* 2011). Notch signalling is important *in vivo* for controlling HSC self renewal and differentiation during hematopoietic stress conditions (Lilly *et al.,* 2011).
- Thrombopoietin (THPO) and angiopoietin (Ang-1), which bind to cell surface receptors MPL and Tie2, respectively, are expressed on HSCs. These cytokines are thought to be important as THPO and Ang-1 knockout mice have decreased numbers or defects in bone marrow HSCs. The interaction of Ang-1 with its Tie2 receptor activates β1-integrin and N-cadherin, enhances quiescence, and maintains the long term repopulating ability of HSCs; it also protects against apoptosis by activating the PI3K pathway (Lilly *et al.,* 2011)
- Osteopontin, a matrix glycoprotein expressed by the osteoblasts supports the adhesion of HSC to the osteoblastic niche and negatively regulates HSC proliferation, contributing to the maintenance of a quiescent state (Guerrouahen *et al.,* 2011).
- The chemokines and their receptors can control HSC behaviour The best understood chemokine, in this regards, is the stromal-derived factor-1 (SDF-1), also called chemokine C-X-C motif ligand 12 (CXCL12). The SDF-1 receptor is the C-X-C chemokine receptor type 4 (CXCR4) and is expressed on HSCs and progenitors (Lilly *et al.,* 2011). SDF-1 belongs to α-chemokines that functions as chemoattractant for both committed and primitive hematopoietic progenitors and regulates embryonic development including organ homeostasis (Ratajczak *et al.,* 2006). SDF-1 counteracts with its cognate receptor CXCR4, that is widely expressed in several tissues including hematopoietic and endothelial cells. SDF-1/CXCR4 signaling plays a critical role during embryonic development by regulating B-cell lymphopoiesis, myelopoiesis in bone marrow and heart ventricular septum formation. In addition, SDF-1 has been shown to mediate the recruitment of endothelial progenitor cells (EPCs) from the bone marrow through a CXCR4 dependent mechanism suggesting the functional role in vasculogenesis in which EPCs could form blood vessels (Chotinantakul & Leeanansaksiri, 2012).
- Both HSC and osteoblasts express N-cadherin, and bone marrow imaging studies suggest that spindle-shaped bone-lining osteoblasts (so-called SNO cells) communicate with HSC through N-cadherin interactions. BrdU incorporation studies demonstrated that quiescent HSCs with moderate N-cadherin reside close to osteoblasts which have high N-cadherin expression. Furthermore, upregulation of N-cadherin on osteoblasts increases adherence of HSC on the endosteal surface, which is associated with HSC quiescence and diminished differentiation (Coskun & Hirschi, 2010).

- Early HSC characterization studies led to the discovery of a growth factor secreted by osteoblasts, called stem cell factor (SCF) that regulates HSC activity *in vivo* and self-renewal *in vitro*. HSCs express a transmembrane receptor tyrosine kinase called stem cell factor receptor (C-Kit) that can bind to SCF, activating intracellular signaling important for HSC regulation (Audet *et al.*, 2002).
- Members of the Wingless (Wnt) family of lipid-modified proteins have been also investigated in hematopoiesis. The Frizzled (Fzd) receptors act as Wnt receptors which activate downstream signaling in the Ctnnb1-dependent canonical and non-canonical pathways. Several components of the Wnt signaling machinery have been shown to play a role in HSC self-renewal. Wnt signaling is important in bone formation and in enlargement of endosteal surfaces. Several lines of evidence suggest that Wnt signaling in endosteal stromal cells may affect HSC maintenance, not by intrinsic signals, but by signals originated from the stromal cells. Wnt signaling may not be intrinsically involved in the maintenance of normal HSC during hemostasis or self-renewal. However, there are data suggesting that changes of Wnt signaling in endosteal stromal cells affect HSC maintenance through extrinsical mechanisms (Renstrom *et al.*, 2010).

Molecule	**Function**
Jagged (Notch receptor)	Control HSC self-renewal and differentiation during hematopoietic stress conditions
Thrombopoietin and angiopoietin	Maintenance of HSC in the niche in quiescence state with a link to cell-cycle control
Osteopontin	Maintenance of a quiescent state
SDF-1 (or CXCL12)	Expressed in the stromal cells; attracts HSC
CXCR-4	Expressed in HSCs; receptor of SDF-1
N-cadherin	HSC quiescence and decrease in differentiation
SCF	Activation of intracellular signaling important for HSC regulation
Wnt	HSC self-renewal

Table 1. Molecules that regulate HSC activity

2.2. Endothelial cells

Endothelial cells were proposed to be important in the HSC microenvironment. *In vivo* and tissue section imaging studies localize HSCs next to endothelial cells. Also, endothelial cells secrete soluble factors that can expand human primitive hematopoietic cells *ex vivo*. However, endothelial cells have not yet been shown to be a necessary regulatory component of the HSC microenvironment *in vivo* (Frisch *et al.*, 2008).

http://dx.doi.org/10.5772/55508

2.3. Osteoclasts

These cells are formed by fusion of multiple granulocyte–macrophage progenitor cells, a process mediated by osteoblasts. It reabsorb the mineralized bone matrix formed by chondrocytes or osteoblasts and located in endosteal niches. The role of osteoclasts in hematopoiesis remains a controversial issue. It has been reported that osteoclasts degrade endosteal niche components and enhance mobilization of hematopoietic progenitor cells (Kollet *et al.*, 2006). Enzymes secreted by osteoclasts are responsible for the release of HSCs from the endosteal niche. These enzymes are able to cleave factors that promote the interaction between HSCs and their niche. On the other hand, results have been published that suggest that osteoclast activity can promote lodgment of HSCs at the endosteal niche (ter Huurner *et al.*, 2010). Further studies will be needed to clarify how osteoclasts regulate hematopoietic stem and progenitor cell behavior (Sugiyama & Nagasawa *et al.*, 2012).

3. Vascular niche

Hematopoiesis and vascularization occur concurrently during development. In fact, HSCs and endothelial cells are derived from the same progenitor cells (termed hemangioblasts) at the embryonic stage and are closely related to the ontogeny of hematopoiesis (Yin & Li, 2006). The presence of a second specialized HSC microenvironment in the bone marrow has recently been postulated, as a large proportion of CD150+ HSCs was observed to be attached to the fenestrated endothelium of bone marrow sinusoids (Wilson & Trumpp, 2006).

The vascular niche promotes proliferation and differentiation, active cycling, and short-term HSCs. Most purified HSCs, containing CD150+ CD48− CD41− Lin− cells, were found to be mainly associated with the sinusoidal endothelium lining blood vessels, suggesting that endothelial cells create a cellular niche for HSCs. Endothelial cells in the vascular niche environment contacting HSCs also provide maintenance signals on the HSC behavior (Can, 2008). However, it is essential to keep in mind that vasculature is not compartmentalized to the central region of bone marrow and, in fact, the endosteal region of bone is also vascularized. Therefore, the proposed osteoblast and vascular niches within marrow are not completely separable, and may function interdependently to generate and sustain HSCs (Coskun & Hirschi, 2010).

The vascular niche has been shown to produce factors important for mobilization, homing, and engraftment of HSC. Endothelial cells expressing vascular cell-adhesion molecule-1 (VCAM-1) associate closely with megakaryocytes and their progenitors through VLA-4 in response to chemotactic factors, stromal cell-derived factor-1 (SDF1) and fibroblast growth factor-4 (FGF4); thus, they provide a niche for megakaryocyte maturation and platelet production (Avecilla *et al.*, 2004).

Two perivascular cell groups that possess mesenchymal cell properties function as niche cells: 1. CXC chemokine ligand 12 (CXCL-12)-abundant reticular cells (CAR cells), and 2. Nestin 234+ mesenchymal stem cells (Nakamura-Ishizu & Suda, 2012).

3.1. CAR cells

In human bone marrow, such reticular cells constitute the subendothelial (adventitial) layer of sinusoidal walls projecting a reticular process that is in close contact with HSCs. Interestingly, these reticular cells were derived from a specific subset of mesenchymal cells (CD146þ) that had been shown to produce either reticular or endosteum of the ectopic hematopoietic microenvironment (HME), referred to as skeletal stem cells (Sugiyama & Nagasawa, 2012).

These cells have recently been shown to be high secretors of SDF-1 (CXCL12), and as a result they have been named CXCL12 abundant reticular (CAR) cells. Phenotypically they express VCAM-1, CD44, 238 platelet-derived growth factor receptor (PDGFRα and PDGFRβ) and possess adipogenic and osteogenic differentiation capacity (Nakamura-Ishizu & Suda, 2012). Histochemical analysis revealed that all bone marrow sinusoidal endothelial cells are surrounded by a proportion of CAR cells, however, CAR cells do not express the pan-endothelial marker platelet/endothelial cell-adhesion molecule 1 (PECAM-1)/CD31 or the smooth muscle cell maker and smooth muscle α-actin (SMαA), suggesting that they are different from endothelial cells and smooth muscle cells (Sugiyama & Nagasawa, 2012).

The depletion of CAR cells using a diphtheria toxin mouse model reduces the cycling of lymphoid and erythroid progenitors, as well as the total HSC cell number and cell cycling, reflecting that CAR cells regulate the proliferation of HSC rather than its quiescence. Ablation of CAR cells did not influence other niche cell compartments such as endothelial cells or osteoblasts (Lilly *et al.*, 2011). However, results based on the short term duration of the CAR cell deletion underestimate the influence that CAR cells may convey on other niches.

3.2. NES+ cells

Nestin is an intermediate filament protein that was originally identified as a marker of neural progenitor. Its expression has subsequently been detected in a wide range of progenitor cells and endothelial cells (Sugiyama & Nagasawa, 2012). NES+ MSCs can be differentiated into adipocytes, osteoblasts and chondrocytes, and their HSC regulatory function is modified by sympathectomy or by treatment with G-CSF (which downregulates HSC ability to express CXCL12, SCF, angiopoietin, IL7, and vascular cell adhesion molecule 1 (VCAM1)) (Wang & Wagers, 2011). Nestin+ cells exhibited multilineage differentiation into various mesenchymal cell lineages including osteoblasts (Nagasawa *et al.*, 2011).

Nestin+ cells express high levels of genes involved in the regulation of HSCs: Cxcl12, c-kit ligand, angiopoietin-1, interleukin-7, vascular cell adhesion molecule-1 and osteopontin. Recently, it was demonstrated that bone marrow CD169+ macrophages are able to maintain the HSCs through CXCL12 levels and through nestin 265+ niche cells, which emphasizes the dense crosstalk among various niches (Nakamura-Ishizu & Suda, 2012).

http://dx.doi.org/10.5772/55508

4. The relationship between the endosteal and vascular niches

It is well known that HSC circulation involves HSCs leaving the bone marrow, entering in the vascular system (mobilization), and returning to the bone marrow (homing). However, the underlying physiological function of these events is not well understood (Yin & Li, 2006).

The endosteal niche, localized at the inner surface of the bone cavity and with abundant osteoblasts, might serve as a reservoir for long-term HSC storage in a quiescent state, whereas the vascular niche, which consists of sinusoidal endothelial cell lining blood vessel, provides an environment for short-term HSC proliferation and differentiation. Both niches act together to maintain hematopoietic homeostasis or to restore it after damage (Guerrouahen *et al.*, 2011).

Based on *in vivo* immunofluorescence with signaling lymphocytic activation molecule (SLAM), a family of cell surface receptors, Kiel & Morrison (2006) identified the vascular niche of HSCs in several tissues, also known as the sinusoidal endothelial niche. Though the two kinds of HSCs niches were anatomically and functionally defined, accumulated data suggests that endosteal and vascular niches overlap in both location and function (Figure 1). Three dimension imaging determined that there are abundant vascular structures on the surfaces of trabecular bones, and that those vessels and endosteal surfaces are intimately coupled with each other within a trabecular region (Wang & Wargers, 2011).

The major difference between both microenvironments is the oxygen level. Higher in the vascular niche than in the endosteal niche under hypoxia, HSCs would move to the vascular niche and resume then the cell cycle in order to restore hematopoiesis. HSCs would then return to the endosteal niche where they would again be maintained in the G0 state (Parmar *et al.*, 2007).

Numerous examples of HSC-endothelial cross-talk exist, although most studies have focused on the function of the endothelial cell in HSC homing. Recent investigations have suggested roles for the vascular CAMs E- and P-selectin and VCAM-1 in HSC homing to bone marrow, as well as for the chemokine SDF-1 (Calmone & Sipkins, 2008).

5. HSCs outside their niches

To turn the situation even more complex, HSCs are not static. Although the vast majority of HSCs in adult humans is located in the bone marrow, HSCs show remarkable mobility. In response to specific signals they can exit and re-enter the endosteal bone-marrow HSC niche, processes known as mobilization and homing, respectively (Wilson *et al.*, 2006).

The use of mobilizing regimens for the collection of HSCs from the peripheral blood of donors rather than from the BM soon became common clinical practice in transplantation settings far before understanding the mechanisms underlying this phenomenon. The most efficient cytokine currently used in the clinical practice to mobilize HSCs is the granulocyte colony-stimulating factor (G-CSF) or its pegylated form (Peg G-CSF) used in a single administration. It was then shown that G-CSF-induced mobilization involved the modulation of the SDF-1/CXCR4 axis, whereby the reduction of the SDF-1 levels and the upregulation of its receptor CXCR4 were correlated with stem cell mobilization. However, although evidence suggested that the mobilization effect of G-CSF lies in its capacity to modify the SDF1 gradient (CXCL12) between the bone marrow and the peripheral blood, favoring the release of HSCs, the exact mechanism by which this occurs has not been completely clarified (Lymperi *et al.*, 2010).

The release of HSCs not only occurs during mobilization but it is also observed during homeostasis, when a small number of HSCs are constantly released into the circulation. Although their precise physiological role remains unclear, they might provide a rapidly accessible source of HSCs to repopulate areas of injured bone marrow (Wilson *et al.*, 2006).

6. HSC niches and disease

The elucidation of the cellular components and molecular effectors of the HSC niche has raised obvious interest on whether analogous regulatory processes are involved in the biology of bone marrow tumors. Increasing evidences point toward critical roles of nonautonomous, microenvironmental factors in the development, progression, and drug resistance of different malignancies evolving in the bone marrow (Carlesso & Cardoso, 2010). Although most hematopoietic malignancies are likely to arise from mutations that inappropriately activate hematopoietic cell proliferation and survival pathways, recent data demonstrate that hematopoiesis can also be dysregulated by alterations in the niche, with defects in HSCs themselves arising secondarily (Wang *et al.*, 2012). The extent of the reliance of these tumors on the microenvironment appears to be dependent upon the type and stage of malignancy. At one extreme is a neoplastic growth that is dependent on dysregulated cell-cell interactions and signalling pathways within the microenvironment. At the other end of the spectrum there are malignancies that exhibit an absolute dependence on normal microenvironmental cues for disease progression, such as the expression of specific cytokines and growth factors (Guerrouahen *et al.*, 2011).

Several studies have provided insights into the role of altered microenvironment signaling leading to myelofibrosis, myeloma, and myelodysplastic syndromes (Noll *et al.*, 2012). Lataillade *et al.* (2008) suggested that an imbalance between endosteal and vascular niches may be important in idiopathic disorder characterized by bone marrow fibrosis (primary myelofibrosis), leading to the development of clonal stem cell proliferation. The most compelling recent example comes from work in which miRNA processing was disrupted in osteoblast precursors and mice developed myelodysplasia, which in rare cases progressed to myeloid leukaemia by 3 weeks of age; notably, HSCs transplanted from these mice into wild-type recipients did not transfer myelodysplasia, indicating that the HSCs were not intrinsically competent to produce pathologic changes (Hosokawa *et al.*, 2010).

http://dx.doi.org/10.5772/55508

6.1. Leukemic Stem Cells

Leukemic stem cells (LSCs) were first described in 1994. The paradigm of cancer stem cells considers leukemia a hierarchical disease process whose growth is sustained by a rare population of LSCs. LSCs maintain the capacity to self-renewal and give rise to malignant progeny with extensive proliferative potential (Flynn & Kaufman, 2007).

It has been speculated that the transformation involves at least a 2-step process, one mutation blocking differentiation and another event conferring a proliferative advantage to its progeny (figure 2). Other LSCs may result from dedifferentiation of more committed progenitors that reacquire the ability of self-renewal prior to accumulating transforming mutations (Blair *et al.*, 1998).

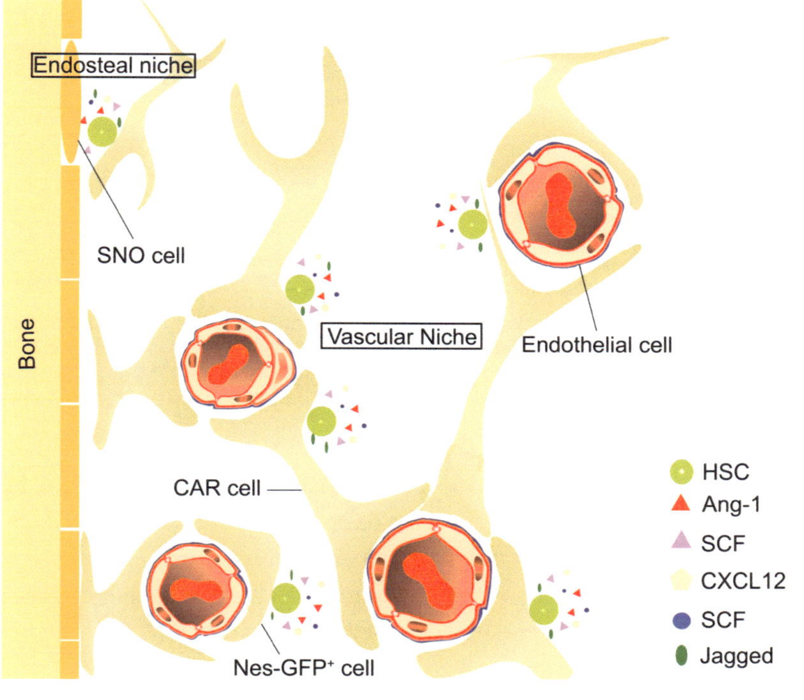

Figure 1. Reticular niches created by mesenchymal progenitors might maintain and regulate HSCs. A model for the localization of HSCs and their association with candidate cell niches in the bone marrow (Modified from Nagasawa *et al.*, 2011).

There is much greater heterogeneity in the phenotype of LSCs than has been previously thought, indicating the inadequacy of the currently used surface antigens or biochemical markers as criteria for LSCs isolation. Evidences from the literature suggest that LSCs are a moving target and its identification depends on many factors including the receptiveness of the murine model used in the experimental design. In addition, it is debatable whether results derived from highly artificial animal models could be extrapolated for the situation in human

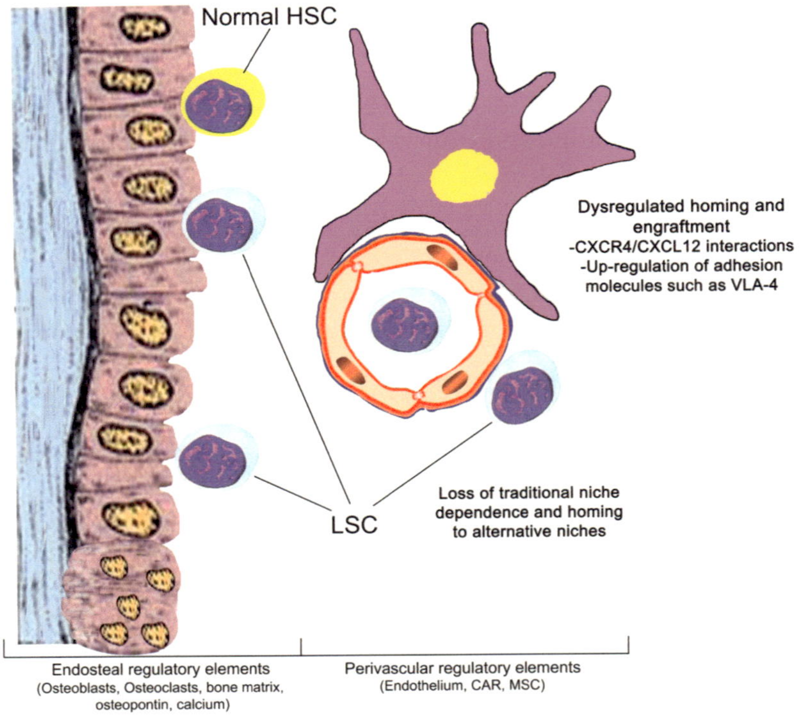

Figure 2. Putative mechanisms for AML stem cell and niche interactions *in vivo*. The niche provides support for self-renewal, quiescence, homing, engraftment, and proliferative potential for HSCs. LSCs may impair the function of the normal HSC niche by direct invasion or secretion of substances such as stem cell factor. LSCs can infiltrate these niches leading to enhanced self-renewal and proliferation, enforced quiescence, and resistance to chemotherapeutic agents. (Modified from Lane *et al.*, 2009).

(Lutz *et al.*, 2012). So far there are only few data analyzing the diversity of LSC in individual patients with acute myeloid leukemia (AML). It has been shown that CD34+CD38– subpopulations are heterogeneous in the distribution of mutations compared to the whole blast population at diagnosis. The situation might be similar to the genetic heterogeneity detected in childhood acute lymphoblastic leukemia (Anderson *et al.*, 2011).

In BCR-ABL–associated leukemias, the transformation occurs at the stem cell and progenitor cell level depending on the phenotype and fusion transcript isoform (Castor *et al.*, 2005). It is a population of highly quiescent HSCs expressing BCR-ABL1 that can be isolated from untreated chronic myeloid leukemia (CML) patients and from imatinib-treated CML patients. These quiescent, non-proliferating CD34+ CML cells have been shown to be resistant to a range of pro-apoptotic stimuli including chemotherapy and tyrosine kinase inhibition with imatinib. The quiescent BCR-ABL1-expressing HSCs can be regarded as LSCs. By way of comparison, proliferating CD34+ progenitors in CML are sensitive to imatinib-induced apoptosis that is significantly mediated by the pro-apoptotic BCL-2 family proteins Bim and Bad (Kuroda *et al.*, 2006; Ng, 2012).

In analogy to normal HSC, LSC also need the marrow niche for their malignant self-renewal and dormant state. Perturbing the binding of LSC to the marrow niche through disruption of the adhesive forces might therefore "mobilize" LSC from their protective environment. Molecules such as CD44, CXCR4, N-cadherin, among others, appear to play significant regulatory roles for HSC and LSC trafficking, signaling and homing to their marrow niche (Sugiyama *et al.*, 2006; Spoo *et al.*, 2007; Teicher & Fricker, 2010; Lutz *et al.*, 2012). LSC may also hijack these pathways in a number of ways, for example, up-regulation of the $\alpha4\beta1$ integrin, VLA-4, which mediates adhesion to fibronectin and VCAM-1. Patients with undetectable VLA-4 levels on leukemic blasts had an excellent response to chemotherapy, perhaps indicating that this pathway may mediate a stromal influence on sensitivity to chemotherapy (Matsunaga *et al.*, 2003). These data suggest that interactions between VLA-4 on leukemic cells and fibronectin expressed on BM stromal cells may modulate chemotherapy sensitivity (Doan & Chute, 2012). Another example is the CD44, the receptor for hyaluronic acid, which is also important for hematopoietic stem and progenitor cell homing to the vascular niche. In both, a mouse model of chronic myelogenous leukemia and a xenograft model of human AML, treatment with an anti-CD44 antibody resulted in LSC mobilization from the niche, LSC differentiation and LSC eradication (Krause *et al.*, 2006).

6.2. Multiple myeloma

A number of pathways and cell types have been shown to affect the behaviour of both HSCs in normal hematopoiesis and the malignant myeloma plasma cells. It is through these regulated interactions with cell populations and signalling pathways within the bone marrow microenvironment that myeloma cells are believed to 'hijack' the normal hematopoietic niche to aid the extensive growth and proliferation of tumour cells (Noll *et al.*, 2012). Multiple myeloma (MM) is characterized by the proliferation of a malignant plasma cell clone, initially located in the bone marrow microenvironment. This illness is unique among hematological malignancies in its capacity to cause great bone destruction, leading to pathologic bone fractures and intractable bone pain. This result is the consequence of an imbalance between osteoblastic and osteoclastic activity induced by MM cells (Corre *et al.*, 2007).

Normal plasma cells are dependent on specific signals from bone marrow stem cells to regulate their differentiation, growth and localization. These same signals are required for myeloma cell growth and survival, supporting the notion that the bone marrow provides a permissive environment for disease development (Degrassi *et al.*, 1993). It is evident that the presence of myeloma cells in the bone marrow modulates the expression of cytokines from stromal cells, which enhances their ability to modify the microenvironment to support malignant growth (Noll *et al.*, 2012).

7. Conclusion

During the past few years, the theoretical concept of a specific stem cell microenvironment that was proposed in the 1960s and 1970s, has finally received greater attention. As mentioned, pinpointing the exact location of the hematopoietic stem cell *in vivo* within the bone marrow is difficult, despite advancements in immunohistochemistry, genetic marking of cells, and *in vivo* imaging. Early studies showed that hematopoietic progenitor and stem cells were highly present near the endosteal bone surface, whereas more mature cells were selectively localized centrally within the bone marrow cavity.

The development of hematologic malignancies may be a multi-step process involving mutations both in the hematopoietic cells and/or in cells present in the supportive microenvironment. Targeting the niche-HSCs, niche-leukemic cells, or niche itself in hematopoietic malignancies is an attractive potential addition to the therapeutic possibilities. The challenge of stem cell biology is to translate the expansion of biological insights into clinically meaningful improvements for patients with hematological malignancies and related disorders.

Author details

Sergio P. Bydlowski, Debora Levy, Jorge M.L. Ruiz and Juliana Pereira

Laboratory of Genetics and Molecular Hematology, Department of Medicine, University of São Paulo School of Medicine, São Paulo/SP,, Brazil

References

[1] Anderson, K, Lutz, C, Van Delft, F. W, Bateman, C. M, Guo, Y, Colman, S. M, Kempski, H, Moorman, A. V, Titley, I, Swansbury, J, Kearney, L, Enver, T, & Greaves, M. Genetic variegation of clonal architecture and propagating cells in leukaemia. Nature. (2011). , 469, 356-361.

[2] Audet, J, Miller, C. L, Eaves, C. J, & Piret, J. M. Common and distinct features of cytokine effects on hematopoietic stem and progenitor cells revealed by dose-response surface analysis. Biotechnol Bioeng. (2002). , 80, 393-404.

[3] Avecilla, S. T, Hattori, K, Heissig, B, Tejada, R, Liao, F, Shido, K, Jin, D. K, Dias, S, Zhang, F, Hartman, T. E, Hackett, N. R, Crystal, R. G, Witte, L, Hicklin, D. J, Bohlen, P, Eaton, D, Lyden, D, De Sauvage, F, & Rafii, S. Chemokine-mediated interaction of hematopoietic progenitors with the bone marrow vascular niche is required for thrombopoiesis. Nat Med. (2004). , 10, 64-71.

[4] Blair, A, Hogge, D. E, & Sutherland, H. J. Most acute myeloid leukemia progenitor cells with long-term proliferative ability in vitro and in vivo have the phenotype CD34/CD71-/HLA-DR-. Blood. (1998). , 92, 4325-4335.

[5] Butler, J. M, Nolan, D. J, Vertes, E. L, Varnum-finney, B, Kobayashi, H, Hooper, A. T, Seandel, M, Shido, K, White, I. A, Kobayashi, M, Witte, L, May, C, Shawber, C, Ki-

mura, Y, Kitajewski, J, Rosenwaks, Z, Bernstein, I. D, & Rafii, S. Endothelial cells are essential for the self-renewal and repopulation of Notch-dependent hematopoietic stem cells. Cell Stem Cell. (2011). , 6, 251-264.

[6] Can, A. Haematopoietic stem cells niches: interrelations between structure and function. Transfus Apher Sci. (2008). , 38, 261-268.

[7] Carlesso, N, & Cardoso, A. A. Stem cell regulatory niches and their role in normal and malignant hematopoiesis. Curr Opin Hematol. (2010). , 17, 281-286.

[8] Castor, A, Nilsson, L, Astrand-grundström, I, Buitenhuis, M, Ramirez, C, Anderson, K, Strömbeck, B, Garwicz, S, Békássy, A. N, Schmiegelow, K, Lausen, B, Hokland, P, Lehmann, S, Juliusson, G, Johansson, B, & Jacobsen, S. E. Distinct patterns of hematopoietic stem cell involvement in acute lymphoblastic leukemia. Nat Med. (2005). , 11, 630-637.

[9] Chotinantakul, K, & Leeanansaksiri, W. Hematopoietic stem cell development, niches, and signaling pathways. Bone Marrow Res. (2012).

[10] Corre, J, Mahtouk, K, Attal, M, Gadelorge, M, Huynh, A, Fleury-cappellesso, S, Danho, C, Laharrague, P, Klein, B, Rème, T, & Bourin, P. Bone marrow mesenchymal stem cells are abnormal in multiple myeloma. Leukemia. (2007). , 21, 1079-1088.

[11] Coskun, S, & Hirschi, K. K. Establishment and regulation of the HSC niche: Roles of osteoblastic and vascular compartments. Birth Defects Res C Embryo Today. (2010). , 90, 229-242.

[12] Degrassi, A, Hilbert, D. M, Rudikoff, S, Anderson, A. O, Potter, M, & Coon, H. G. In vitro culture of primary plasmacytomas requires stromal cell feeder layers. Proc Natl Acad Sci USA. (1993). , 90, 2060-2064.

[13] Doan, P. L, & Chute, J. P. The vascular niche: home for normal and malignant hematopoietic stem cells. Leukemia. (2012). , 26, 54-62.

[14] Flynn, C. M, & Kaufman, D. S. Donor cell leukemia: insight into cancer stem cells and the stem cell niche. Blood. (2007). , 109, 2688-2692.

[15] Frisch, B. J, Porter, R. L, & Calvi, L. M. Hematopoietic niche and bone meet. Curr Opin Support Palliat Care. (2008). , 2, 211-217.

[16] Guerrouahen, B. S, Al-hijji, I, & Tabrizi, A. R. Osteoblastic and vascular endothelial niches, their control on normal hematopoietic stem cells, and their consequences on the development of leukemia. Stem Cells Int. (2011).

[17] Hosokawa, K, Arai, F, Yoshihara, H, Iwasaki, H, Hembree, M, Yin, T, Nakamura, Y, Gomei, Y, Takubo, K, Shiama, H, Matsuoka, S, Li, L, & Suda, T. Cadherin-based adhesion is a potential target for niche manipulation to protect hematopoietic stem cells in adult bone marrow. Cell Stem Cell. (2010). , 6, 194-198.

[18] Kiel, M. J, & Morrison, S. J. Maintaining hematopoietic stem cells in the vascular niche. Immunity. (2006). , 25, 862-864.

[19] Kollet, O, Dar, A, Shivtiel, S, Kalinkovich, A, Lapid, K, Sztainberg, Y, Tesio, M, Samstein, R. M, Goichberg, P, Spiegel, A, Elson, A, & Lapidot, T. Osteoclasts degrade endosteal components and promote mobilization of hematopoietic progenitor cells. Nat. Med. (2006). , 12, 657-664.

[20] Krause, D. S, Lazarides, K, Von Andrian, U. H, & Van Etten, R. A. Requirement for CD44 in homing and engraftment of BCR-ABL-expressing leukemic stem cells. Nat Med. (2006). , 12, 1175-1180.

[21] Kuroda, J, Puthalakath, H, Cragg, M. S, et al. Bim and Bad mediate imatinib-induced killing of Bcr/Abl+ leukemic cells, and resistance due to their loss is overcome by a BH3 mimetic. Proc Natl Acad Sci USA (2006). , 103, 14907-14912.

[22] Lane, S. W, Scadden, D. T, & Gilliland, D. G. The leukemic stem cell niche: current concepts and therapeutic opportunities. Blood. (2009). , 114, 1150-1157.

[23] Lataillade, J. J, Pierre-louis, O, Hasselbalch, H. C, Uzan, G, Jasmin, C, & Martyré, M. C. Le Bousse-Kerdilès MC; French INSERM and the European EUMNET Networks on Myelofibrosis. Does primary myelofibrosis involve a defective stem cell niche? From concept to evidence. Blood. (2008). , 112, 3026-3035.

[24] Lilly, A. J, Johnson, W. E, & Bunce, C. M. The haematopoietic stem cell niche: new insights into the mechanisms regulating haematopoietic stem cell behaviour. Stem Cells Int. (2011).

[25] Lutz, C, Hoang, V. T, Buss, E, & Ho, A. D. Identifying leukemia stem cells- Is it feasible and does it matter? Cancer Lett. (2012). Epub ahead of print].

[26] Lymperi, S, Ferraro, F, & Scadden, D. T. The HSC niche concept has turned 31. Has our knowledge matured? Ann N Y Acad Sci. (2010). , 1192, 12-18.

[27] Matsunaga, T, Takemoto, N, Sato, T, Takimoto, R, Tanaka, I, Fujimi, A, Akiyama, T, Kuroda, H, Kawano, Y, Kobune, M, Kato, J, Hirayama, Y, Sakamaki, S, Kohda, K, Miyake, K, & Niitsu, Y. Interaction between leukemic-cell VLA-4 and stromal fibronectin is a decisive factor for minimal residual disease of acute myelogenous leukemia. Nat Med. (2003). , 9, 1158-1165.

[28] Nagasawa, T, Omatsu, Y, & Sugiyama, T. Control of hematopoietic stem cells by the bone marrow stromal niche: the role of reticular cells. Trends Immunol. (2011). , 32, 315-320.

[29] Nakamura-ishizu, A, & Suda, T. Hematopoietic stem cell niche: An interplay among a repertoire of multiple functional niches. Biochim Biophys Acta. (2012). Epub ahead of print].

[30] Noll, J. E, Williams, S. A, Purton, L. E, & Zannettino, A. C. Tug of war in the haematopoietic stem cell niche: do myeloma plasma cells compete for the HSC niche? Blood Cancer J. (2012). e91.

[31] Oh, I. H, & Kwon, K. R. Concise review: multiple niches for hematopoietic stem cell regulations. Stem Cells. (2010). , 28, 1243-1249.

[32] Parmar, K, Mauch, P, Vergilio, J. A, Sackstein, R, & Down, J. D. Distribution of hematopoietic stem cells in the bone marrow according to regional hypoxia. Proc Natl Acad Sci USA. (2007). , 104, 5431-5436.

[33] Ratajczak, M. Z, Zuba-surma, E, Kucia, M, Reca, R, Wojakowski, W, & Ratajczak, J. The pleiotropic effects of the SDF-1-CXCR4 axis in organogenesis, regeneration and tumorigenesis. Leukemia. (2006). , 20, 1915-1924.

[34] Renström, J, Kröger, M, Peschel, C, & Oostendorp, R. A. How the niche regulates hematopoietic stem cells. Chem Biol Interact. (2010). , 184, 7-15.

[35] Schofield, R. The relationship between the spleen colony-forming cell and the haemopoietic stem cell. Blood Cells. (1978). , 4, 7-25.

[36] Shen, Y, & Nilsson, S. K. Bone, microenvironment and hematopoiesis. Curr Opin Hematol. (2012). , 19, 250-255.

[37] Spoo, A. C, Lübbert, M, Wierda, W. G, & Burger, J. A. CXCR4 is a prognostic marker in acute myelogenous leukemia. Blood. (2007). , 109, 786-791.

[38] Sugiyama, T, Kohara, H, Noda, M, & Nagasawa, T. Maintenance of the hematopoietic stem cell pool by CXCL12-CXCR4 chemokine signaling in bone marrow stromal cell niches. Immunity. (2006). , 25, 977-988.

[39] Sugiyama, T, & Nagasawa, T. Bone marrow niches for hematopoietic stem cells and immune cells. Inflamm Allergy Drug Targets. (2012). , 11, 201-206.

[40] Teicher, B. A, & Fricker, S. P. CXCL12 (SDF-1)/CXCR4 pathway in cancer. Clin Cancer Res. (2010). , 16, 2927-2931.

[41] ter Huurne MFigdor CG, Torensma R. Hematopoietic stem cells are coordinated by the molecular cues of the endosteal niche. Stem Cells Dev. (2010). , 19, 1131-1141.

[42] Wang, H, Zhang, P, Liu, L, & Zou, L. Hierarchical organization and regulation of the hematopoietic stem cell osteoblastic niche. Crit Rev Oncol Hematol. (2012). Epub ahead of print]

[43] Wang, L. D, & Wagers, A. J. Dynamic niches in the origination and differentiation of haematopoietic stem cells. Nat Rev Mol Cell Biol. (2011). , 12, 643-655.

[44] Wilson, A, & Trumpp, A. Bone-marrow haematopoietic-stem-cell niches. Nat Rev Immunol. (2006). , 6, 93-106.

[45] Yin, T, & Li, L. The stem cell niches in bone. J Clin Invest. (2006). , 116, 1195-1201.

Chapter 3

Expression Profile of Galectins (Gal-1, Gal-9, Gal-11 and Gal-13) in Human Bone Marrow Derived Mesenchymal Stem Cells in Different Culture Mediums

Faouzi Jenhani

Additional information is available at the end of the chapter

http://dx.doi.org/10.5772/55383

1. Introduction

Mesenchymal stem cells (MCSs) are refined as undifferencitaed cells that are capable of self renewal and differentiation into several cell types such chondrocyte, adipocyte osteocyte, myocyte and neuron-like cells. MSC can be isolated from bone marrow umbilical cord blood adipose tissue placenta. Although bone marrow(BM) has been regarded as a major source of MSC umbilical cord blood has been regarded as an alternative source for isolation of MSC. Human umbilical cord blood derived mesenchymal stem cell (hUCB-MSCs) have a capacity similar to that of BM-MCSs for multi-lineage differentiation

Researchers are interested in these cells because of their ability of differentiating into multiple mesenchymal and non mesenchymal lineages. Furthermore, MSCs evoke only minimal immunoreactivity and they display trophic, anti-inflammatory, and immunomodulatory capacities, through secretion of bioactive soluble factors with anti-inflammatory and immunomodulatory effects in vivo. These properties were confirmed by both experimental and clinical studies which demonstrated that the MSCs support hematopoiesis and enhance the engraftment of hematopoietic stem cells (HSC) after cotransplantation, which may contribute to a reduced incidence of graft versus host disease (GVHD). [1,2,3,4,5,6,7]

The galectins are a family of soluble lectins characterized by their affinity for b-galactoside residues. These proteins have recently attracted increasing attention because of theirinvolvement in various physiological and pathological processes. In addition, these proteins have recently attracted increasing attention of cancer biologists because of their essential functions including development, differentiation, cell–cell adhesion, cell–matrix interaction, growth regulation, apoptosis, RNA splicing, and tumor metastasis. Also, it has been shown that galectins' levels are altered in various cancers. [8,9,10]

Galectins-1,3,9 and 13 were among the best characterized members of this family. These galactins possesses several functions and were expressed in many tissues.

Given the importance of galectins, we investigated in this work, their expression by BM MSCs in different culture mediums

2. Materiel and methods

2.1. Preparation of hBM-MSC

MSCs were extracted and isolated from two sources adult the bone marrow. H- *BM cells* was harvested from the sternum or the iliac crest of healthy donors [20] after obtaining an informed consent was obtained in the Bone Marrow Graft National Center of Tunisia. The mean age of donors was 30 ± 2 years (range, 15-40 years).

MSCs were isolated using the classical plastic adhesion method. Mononuclear cells (MNCs) were isolated from hBM by density gradient (Ficoll Hypaque solution d=1. 077) centrifugation. After centrifugation at 800 g for 20 min at room temperature. MNC layer was removed from the interphase and washed twice with Hank's buffered salt solution and seeded into uncoated T25 or T75 flasks (Becton Dickinson, Bedford, MA, USA) at a cell concentration of $1{\times}10^5$ cells/cm^2 for BM cells and cultured in three condition mediums:

i. *(M1) : basic growth* medium consisting of alpha-Minimum Essential Mediums (alpha -MEM, Gibco BRL, Grand Island, NY, USA), supplemented with 10 % (v/v) fetal bovine serum (Sigma-Aldrich, Bornem, Belgium),100U/mL penicillin, 0. 1 mg/mL streptomycin (Gibco BRL), 2Mm L-glutamine (Gibco BRL), 0. 025 mg/mL fungizone (Gibco BRL) 10% (v/v) of prescreened fetal bovine serum (FBS; Perbio Hyclone, Logan, Utah, USA) and 1 ng/mL fibroblast growth factor2 (FGF2) (AbCys, Paris, France)

ii. (M2): basic growth medium with fetal bovine serum and FGF2 replaced by 5 % (v/v) human patelet lysed (HPL)

iii. (M3): basic growth medium with fetal bovine serum and FGF2 replaced by 10 % (v/v) h (HPL). Cell cultures were incubated at 37°C in a 5% CO2 humidified atmosphere. The medium was changed twice weekly thereafter. When reaching 60% -80% confluence, the adherent cells were detached after treatment with 0. 05% (v/v) trypsin/ 1 mM EDTA solution (Gibco BRL) and expanded by replating at a lower density at 10^3 cells per cm^2. All the studies were performed after 2 passages (P2).

2.2. Induction of the differentiation directions

We induced the differentiation of the MSCs when they reached the confluence. We used three differentiation mediums according to the focused differentiation line:

2.2.1. Osteogenic induction (O)for analysis galectins expression

At 50% confluence, the cells were cultured for 14 days in DMEM-HG (High Glucose 4. 5 g/L) containing 2% FBS, 0. 1 μM dexamethasone (Sigma), 2 mM β-glycerolphosphate (Sigma), and 100 μM ascorbate-2-phosphate (Sigma) with medium changes every 3 days. After 2 weeks of induction, the cells were stained according to Von Kossa's and Alizarin Red methods to detect the presence of calcium deposition into osteocytes.

2.2.2. Adipogenic induction (A) for analysis galectins expression

The adipogenic induction medium consisted of DMEM-LG (Low Glucose 1g/L) supplemented with 20% FBS, 1 μM/L dexamethasone, 0. 5 mmol/L isobutylmethylxanthine (IBMX; Invitrogen), and 60 μmol/L indomethacin (Sigma). Adipogenic differentiation was evaluated after 2 weeks of induction, by the cellular accumulation of neutral lipid vacuoles that were stained with Nile Red (Sigma) and observed by fluorescent microscopy.

2.2.3. Vascular smooth muscle induction (V) for analysis galectins expression

Vascular smooth muscle (VSM) differentiation was obtained in a Mc Coy'5A medium supplemented with 12. 5% FBS, 12. 5% HS (Horse Serum, Invitrogen), 20 μM L-glutamine, 0,8 mM L-serine, 0,15 mM L-asparagine, 1 Mm sodium pyruvate, 5 mM sodium bicarbonate, 1 μM hydrocortisone and amphotericin B and antibodies. The medium was changed every 4 days. The VSM differentiation was evaluated after 3 weeks of induction.

2.3. Immunophenotyping of mesenchymal stem cells by flow cytometry

MSCs were immunophenotypically characterized by flow cytometry using the following fluorochrome (FITC PE, PerCP) marked monoclonal antibodies anti : CD45, CD105, CD106, CD90,CD49, CD34, et CD73 and CD14 (Becton Dickinson and Company BD Biosciences San Jose CA).

2.4. Detection of galectins (Gal-1, Gal-9 Gal-11 and Gal-13), by flow cytometry

To reveal the expression of the galactin we used biotynaled antibodies against human recombinant galectin (Gal-1, Gal-9, Gal-11 and Gal-13). Thus we incubated 1,5 10^6 MSCs cells with the antibody which is specific to the focused galectin, for 30 minutes, at 4° C. We incubated also the MSCs with a control antibody which doesn't recognize any protein, but has the same isotype to evaluate a background noise which correspond to non-specific fixation of the primary antibody.

The MSCs were washed 2 times with PBS 1X solution, then incubated for 30 minutes with Streptavidin coupled with a phycoerythrin (PE) fluorochrome phycoerythrin (PE). Flow cytometry analysis was performed on a FACS calibur, and data were analyzed using CellQuest software (Becton–Dickinson, BD Biosciences San Jose, CA).

2.5. Galectins immunofluorescence detection

MSCs were fixed using 4% Methanol for 30 minutes at +4°C and washed with PBS. The primary antibodies against many anti-antibodies: Gal-1-Streptavidine, Gal-3-Streptavidine, Gal-9-Streptavidin and Gal-13 Streptavidin were diluted in 10% BSA and 0. 2% Tween 20 (1:100) and were incubated at 4°C for 12h followed by washings 3 times with PBS. For immunofluorescence non-specific binding sites were blocked with 10% BSA-PBS. For secondary immunofluorecence Biotine –fluoroscein were diluted in 10% BSA-PBS and incubated for 60 min at +4°C in the dark. Images were taken with camera and with inverse fluorescence microscopy

2.6. RNA isolation and reverse transcription – Polymerase chain reaction before and after induction

2.6.1. RNA isolation

For RT-PCR, Total RNA was extracted from cell culture in total confluence [90%) using TRIZOL reagent (Gibco BRL) according to the manufacturer's instructions. Reverse transcription was carried out using PrimeScript™RTase (Takara: Japan and Applied Biosystems) and the cDNA fragments were amplified using RNase Inhibitor (Takara: Japan and Applied Biosystems): For each reaction, we mixed 5 µl of ARN or 5µl of distilled water (for the control), 6 µl desoxyribonucleotides (dNTP)(TaKaRa Japan and Applied Biosystems), 1µl primers specific of the gene sequence (table 1) et 0. 5 µl of enzyme la Taq polymerase enzyme (Takara: Japan and Applied Biosystems).

2.6.2. RT-PCR before and after induction for the confirmation of the three directions of differentiation

After denaturation at 65°C for 5 min, amplification was carried out by 30 cycles at 30°C for 10 min, 42°C for 60 min and 95°C for 5 min. Primers used are shown in Table 1. The expression of the following genes were analyzed before (D0) and after differentiation (D14) into three directions of differentiation: osteogenesis (O), adipogenesis (A) and vascular smooth muscle (V). Glyceraldehyde 3-phosphate dehydrogenase (GAPDH) gene expression was used as control. The sequences of the oligonucleotides were reported in Table 1. Thermocycling was performed with a gradient thermocycler (Takara, Japan and Applied Biosystems). The analyse of the PCR products was carried out in an electrophoresis gel 1% (Sigma).

Construct	Sequence forward	Sequence reverse
GAPDH	5'- AATCCCATCACCATCTTCCAGG-3'	5'-AGAGGCAGGGATGATGTTCTGG-3'
PAL	5'-CTGGACCTCGTTGACACCTG-3'	5'-GACATTCTCTCGTTCACCGC-3'
Runx2	5'-AACTTCCTGTGCTCGGTGCTG-3'	5'-GGGGAGGATTTGTGAAGACGG-3'
LPL	5'-AAAGCCCTGCTCGTGCTGAC-3'	5'-TAAACCGGGCCACATCCTGT-3'
PPRγ	5'-GGAGAAGCTGTTGGCGGAGA-3'	5'-TCAAGGAGGCCAGCATTGTG-3'
ASMA	5'-TCATGATGCTGTTGTAGGTGGT-3'	5'-CTGTTCCAGCCATCCTTCAT-3'

Table 1. Sequences of the primers used in RT-PCR study

2.6.3. Reverse transcription – Polymerase chain reaction before and after induction for Galectin1,Gal-9,Gal-11 and Gal-13 expression

After denaturation at 65°C for 5 min, amplification was carried out by 30 cycles at 30°C for 10 min, 42°C for 60 min and 95°C for 5 min. Primers used are shown in Table 2.

Construct	Sequence forward	Sequence reverse
GAPDH	5'AATCCCATCACCATCTTCCAGG3'	5'AGAGGCAGGGATGATGTTCTGG3'
Gal 1	5'ATGGCTTGTGGTCTGGTC3'	5'TCAGTCAAAGGCCACACA3'
Gal 9	5'ATGGCCTTCAGCGGTTCC3'	5'CTATGTCTGCACATGGGTCAG3'
Gal 11	5'ATGAGTCAGCCCAGTGGG3'	5'TCAGGAGTGGACACAGTAGAG3'
Gal 13	5'ATGTCCCTGACCCACAG3'	5'TCAATCGCTGATAAGCACT3'

Table 2. Sequences of the primers used in RT-PCR study

Thermocycling was performed with a gradient thermocycler (Takara, Japan and Applied Biosystems). The analyse of the PCR products was carried out in an electrophoresis gel 1% (Sigma).

2.7. Statistical analysis

Data are expressed as mean ± standard error of the mean. Statistical comparisons were performed using the t-test student or Mann Whitney test with the program Graphpad, version. 5. A one-way analysis of variance (ANOVA) was done for paired samples; $p < 0.\ 05$ was considered to be statistically significant.

3. Results

3.1. Morphologic analysis of MSCs derived from bone marrow in terms of culture mediums

To prepare MSCs cultures, we isolated mononuclear cells (MNCs) from bone marrow (hBM). The MNCs derived from bone marrow were put in a density of $1.\ 10^6$ cells / cm^2.

Adherent cell populations from the MNC fraction of hBM samples were generated by expansion culture using 3 different media: (M1) Alpha MEM with 10% FBS and 1 ng/mL FGF_2, (M2) Alpha MEM with 5% HPL, (M3) Alpha with 10% HPL. After two weeks of culture, an adherent and stable cell layer was obtained from BM derived MNC with all medium Mean time for the primary culture to reach subconfluence was 15 days. After one passage(P1), adherent cells displayed a fibroblast-like morphology in culture plate (n=20). The figure 1 showed a particular morphology of cultured MSC.

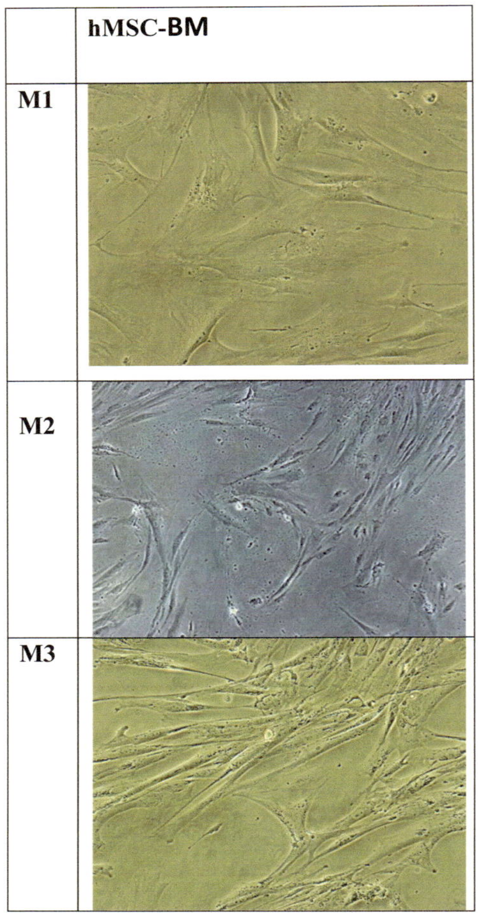

Figure 1. Morphological characterization of hMSC-BM cultured in different media in the P2 (passage): M1: (10% FBS +FGF2), M2:(5% HPL), M3:(10% HPL).

http://dx.doi.org/10.5772/55383

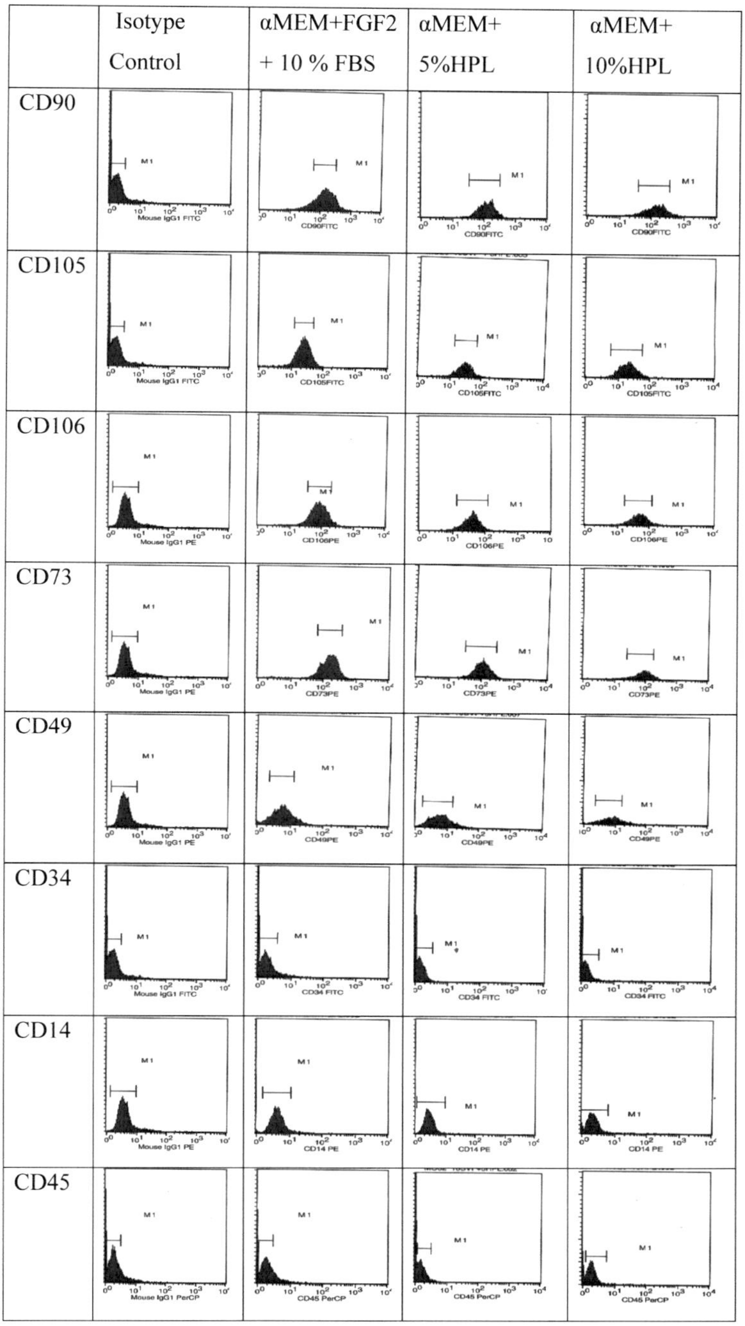

Figure 2. Representative flow cytometry analysis of BM-MSCs.Comparison of membrane antigen expression of BM-MSCs cultured at P2 in different media: M1 :(10% FBS+FGF_2) M2: (5% HPL), M3: (10% HPL)

Also MSCs were immunophenotypically characterized by flow cytometry. This analysis revealed that MSCs were uniformly positive for CD73 CD90 CD105 CD106 but negative for CD14 CD34 and CD45 (figure 2).

The influence of HPL or FBS on the osteogenic and adipogenic differentiation potential of MSCs was investigated after appropriate induction at the second passage (P2). Cells grown in HPL or FBS deposited an extensive mineralized matrix when cultured for 2 weeks in an osteogenic medium [2% FBS), as demonstrated by strong alizarin red staining. These cells also efficiently differentiated into the adipogenic lineage, as indicated by Oil Red O staining of lipid droplets in the cytoplasm following culture in an adipogenic medium.

All populations of BM-MSCs cultured in different conditions with the HPL showed osteogenic, Vascular smooth muscle and chondrogenic differentiation capacity. The difference appears in adipogenic differentiation with 10% HPL (Fig 3).

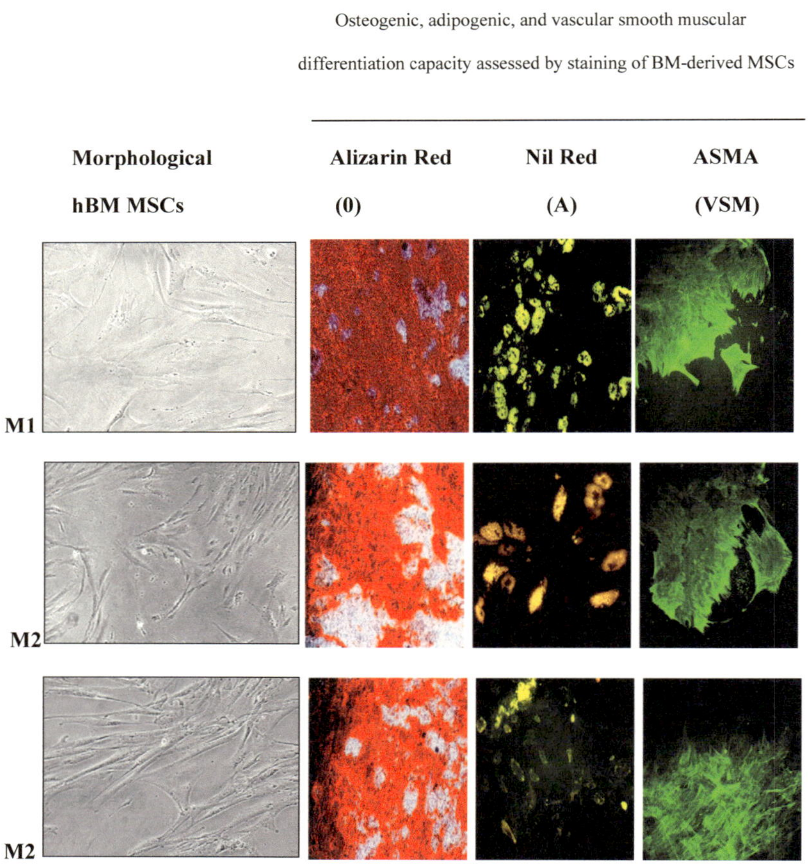

Figure 3. Morphological characterization of hBM- MSCs derived cultured in different media M1, M2 and M3 at P2 and Osteogenic, chondrogenic, adipogenic, and vascular smooth muscular differentiation capacity assessed by staining

To confirm their differentiation potential, after P1 MSCs were plated in specific induction mediums to generate for adipocytes, osteoblasts, or chondrocytes.

To select the differentiated MSCs, we used the RT-PCR to search markers specific for the three differentiation directions. These markers are the alkalin phosphatase (PAL) as osteoblast specific marker the lipoprotein lipase (LPL)as specific of the adipocyte line and the actin-ASMA specific of vascular smooth muscles line. The positive control is the GAPDH. These markers were not expressed before the induction of the differentiation and presented an overexpression after the 14th day from the induction of specific differentiation in the three expansion mediums (M1, M2, et M3) demonstrating the multipotent nature of the MSCs (figure 4).

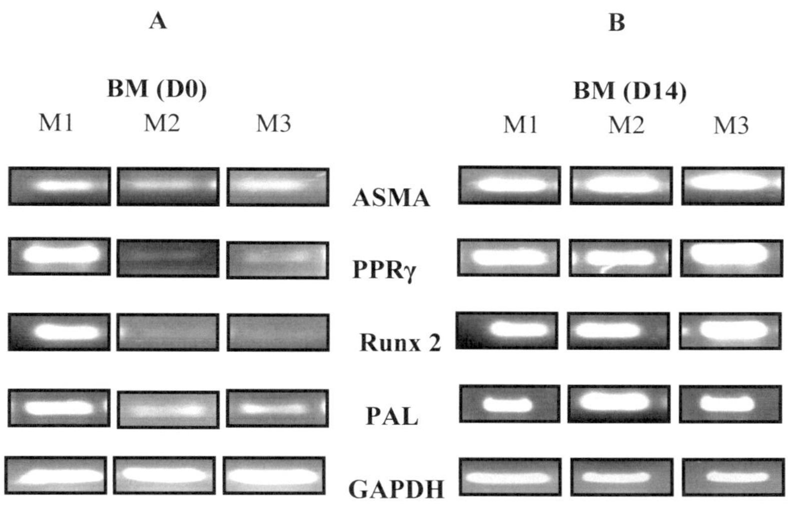

Figure 4. RT-PCR analyses of vascular smooth muscle (ASMA), adipogenic (PPRγ), and osteogenic (Runx2, PAL) markers prior to (D0) (**A**) and after 14 days of differentiation induction (D14) (**B**) of BM-derived MSCs previously cultured in expansion media (M1, M2, M3)

3.2. Detection of galectins in MSCs in terms of the culture mediums by flow cytometry

MSCs from different BM samples were characterized by flow cytometry with a panel of biotynaled antibodies against human recombinant galectin (Gal-1, Gal-9, Gal-11 and Gal-13) at P2 after culture in the three mediums (M1, M2, et M3). Flow cytometry analysis revealed that MSC derived-hBM constitutively expressed galectins (Gal-1, Gal-9, Gal-11, Gal-13) at different mediums but with unequal percentages. We noticed that galectins 9 and 11 are more expressed by cells derived from hBM with an average of [88. 1 ± 1. 8 %) in M1 Meduim in comparison to M2 (16,8 ± 3. 9 %) and M3 (11,6 ± 3. 9%) with $p < 0.01$.

We noticed also that galectin 1 is expressed with in average of (91. 2 ± 1. 4 %) at P2 after culture in the three mediums (M1,M2,M3). In the contrary no expression of galactin 13 have been noted in any medium. (figure 5)

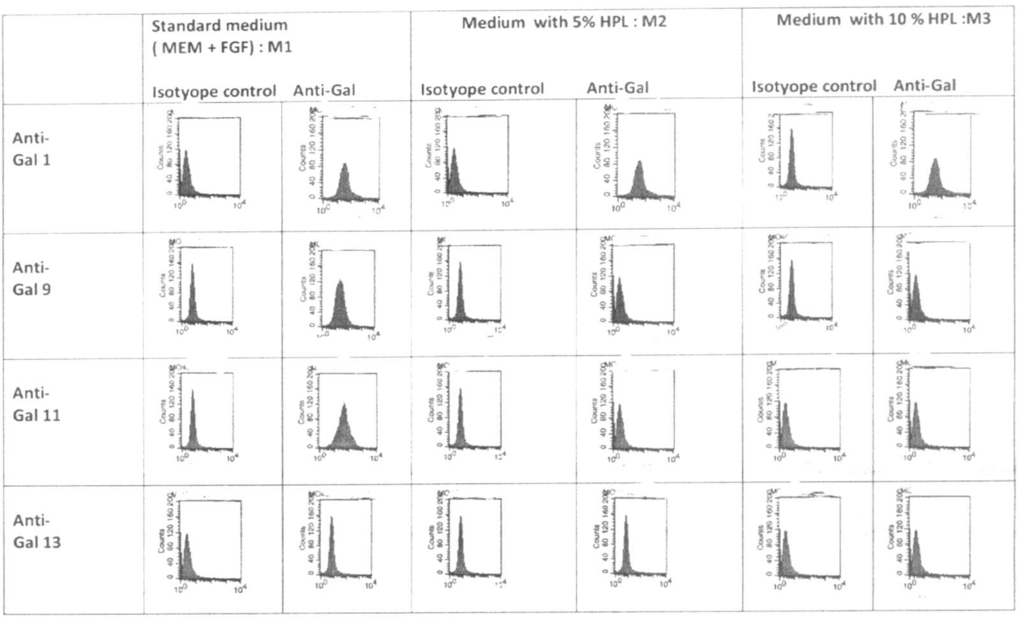

Figure 5. Galectins Gal-1, Gal-9,Gal-11 and Gal-13 antigen expression of BM-MSCs cultured at P2 in different media: M1 (10% FBS+FGF2) M2(5% HPL), M3 (10% HPL) by flow cytometry

3.2.1. Expression of galectins of MSC-BM by microscopy immunofluorecence

Immunofluorecsence of galectin-1 in MSC of normal adult bone marrow was showed expression in three mediums M1,M2,M3. Following this technique, we are also identified galectins gal-9 and gal-11 expression but only in standard medium M1. No expression of gal- 13 have been reveled by hBM-CSM cultured in all media M1,M2 and M3. we are determined the localization of galectin mostly in nucleus cells. (Fig6).

http://dx.doi.org/10.5772/55383

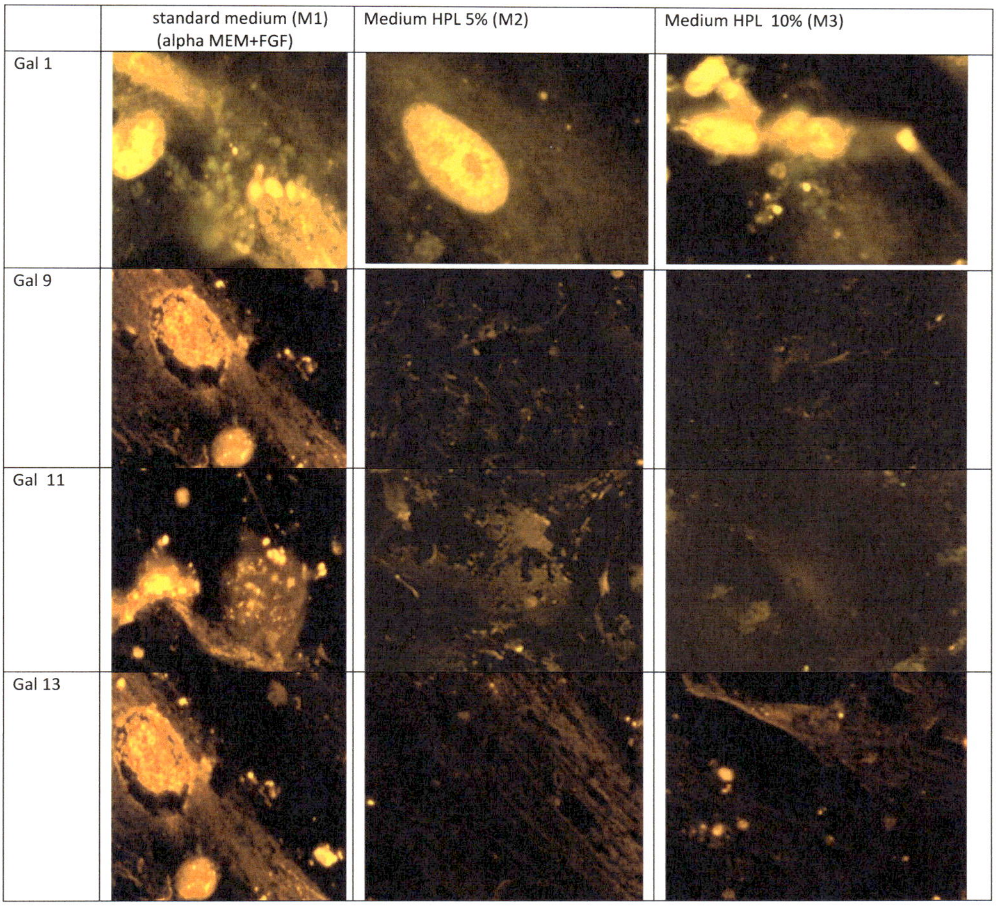

Figure 6. Immunofluorescence Galectin-1, Galectin-9, Galectin -11 and Galactin-13 expression by cultured at P2 in different media: M1 (10% FBS+FGF$_2$) M2 (5% HPL), M3 (10% HPL)

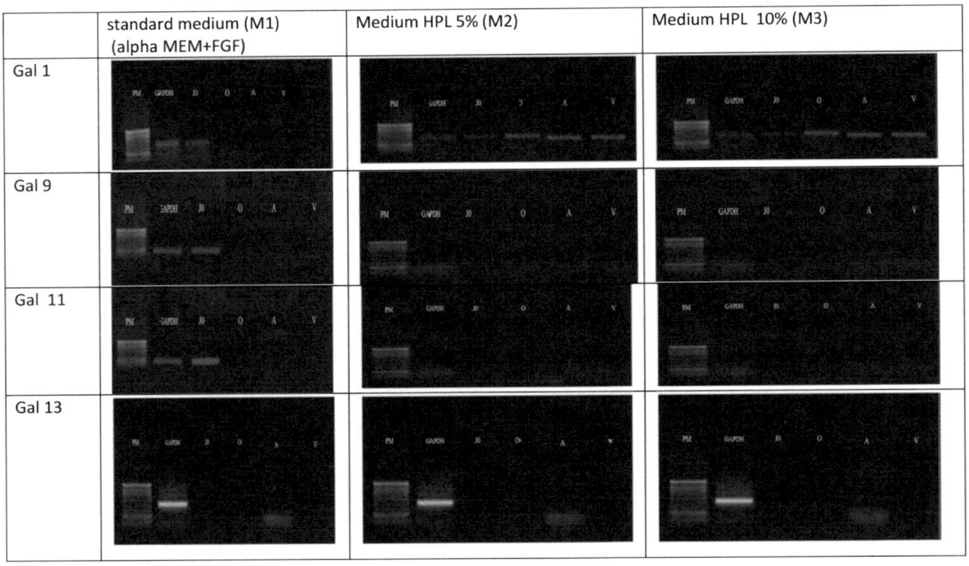

Figure 7. RT –PCR Galectin-1, Galectin-9, Galectin -11 and Galactin-13 expression by cultured at P2 in different media: M1 (10% FBS+FGF$_2$) M2 (5% HPL), M3 (10% HPL) and after differentiation process (O: osteogenic induction, A: Adipogenic induction,V: Vascular Smooth Muscle induction, GAPDH : Glyceraldheyde 3-phospate dehydrogenase, PM: Weight molecular)

3.3. RT –PCR galectin-1,Gal-9,Gal-11 and Gal-13 expression by hBM-MSCs

By RT-PCR, we evaluated the messenger RNA expression of Galectin-1,Galectin-9, Galectin-11 and Galectin -13 in hBM-MSCs before and after differentiation to the three direction : osteoblastic [0), adipocyte (A) and to vascular smooth muscle (VSM).

We had a positive results for hBM-MSCs both before their differentiation and after the induction of the three differentiation directions:

- before the induction of the differentiation process :Gal-1 was expressed by hBM- MSCs in the standard medium M1 (MEM+ FGF2) and also in the three mediums (M2, M3, et M4) which contain the human platelet lysat (HPL). While, after the induction of the MSCs differentiation, Gal-1 was only expressed in MSCs cultivated in mediums containing the HPL (M2, M3), in the three differentiation directions (O, A, VML)
- Concerning the expression of Gal-9 by BM-MSCs: we observed that Gal-9 was expressed by undifferentiated BM-MSCs cultivated in the standard medium (M1: MEM+ FGF2), but absent in MSCs cultivated in mediums containing the HPL (M2, M3). However, after inducing the differentiation process, this galectin was absent in MSCs of the standard medium and had a very low expression in mediums containing the HPL (M2, M3).
- Regarding the expression of Gal-11 by BM-MSCs, we found that Gal- 11 was expressed by undifferenciated BM-MSCs, in the standard medium M1 (MEM+ FGF2) but absent in BM-

MSCs cultivated in M2, M3, and M4. And After the MSCs differentiation Gal-11 was no more expressed in the standard culture's medium (M1 : MEM+ FGF2), while MSCs cultivated in mediums containing the HPL (M2, M3) expressed the Gal- 11 in adipocytes (A) and vascular smooth muscle cells (VML). As to Gal-13, we reported that this galectin wasn't expressed by the undiferenciated MSCs neither in the standard medium nor in the mediums containing the HPL (M1, M2, and M3). Whereas, after the MSCs differentiation, Gal-13 was only expressed by adipocytes. (figure 5)

4. Discussion

Multipotent mesenchymal stromal cells (MSCs) have generated a great debate as a potential source of cells for cell-based therapeutic strategies primarily owing to their intrinsic ability to self renew and differentiate into functional cell types that constitute the tissue in which they exist and also because MSCs express a large number of molecules including the galectins. [11,12]

The galectins are a family of soluble lectins characterized by their affinity for β-galactoside residues. In recent years, galectins have become a major focus of investigation because of their involvement in various physiological and pathological processes. [13,14]

Given that galectins have been shown to have important effects [13,14] we decided to evaluate the expression of galectins (Gal-1, Gal-3, Gal-9 and Gal-13) by MSCs.

In this study we analyzed MSCs cultured and expanded in three conditions mediums: *basic growth* medium consisting of alpha-Minimum Essential Medium in which HPL replaced FBS in order to investigate their morphology, plasticity, proliferation differentiation to osteoblasts, adipocytes, and vascular smooth muscles and their capacity to express galectins in terms of the mediums' compositions. [15,16]

In our laboratory MSCs were characterized and defined according to the International Society for Cellular Therapy minimal criteria. [17,18]. So first, we demonstrated that BM-MSC adhered to plastic. Second as measured by flow cytometry all of MSC derived from BM-MSC were nearly 100% positive for the makers CD90 CD105, and CD73 (≥95%), while they lack the expression of CD45 CD34, and CD14 (≤ 2%). Third, BM-MSC differentiated to osteoblasts adipocytes and vascular smooth cells.

First, we investigated the effect of culture mediums on the adherence and the proliferation of MSCs. Then we demonstrated that the three mediums allow the adherence of MSCs to plastic and their proliferation, but the percentage of confluence and the times it takes to get it, change in terms of the composition of the culture medium. So, the culture mediums containing 5% HPL induced the best MSCs confluence. Thus the most important confluence of the bone marrow's MSCs was reached at the first passage (P1) in the culture's medium 5% HPL after 30 days from their implantation. So, we confirmed that HPL-expanded MSCs show a significantly higher proliferation rate, in comparison with MSCs expanded in the standard medium. So, we can use the HPL as a substitute of FBS in mediums specific to the culture of MSCs destined for clinical applications, to avoid a possible xenogenic contamination caused by the FBS. [19]

The expansion-promoting effect of the HPL is likely to result from the high concentration of natural growth factors that HPL contains. In fact platelet granules contain many growth factors, including platelet-derived growth factor (PDGF), fibroblast growth factor (FGF), insulin-like growth factor (IGF), transforming growth factor (TGFb), platelet factor 4 (PF-4), and platelet-derived epidermal growth factor (PDEGF) [20,21]. These growth factors, which are released from platelet lysate, have been shown to enhance MSC expansion in vitro. [20, 21,22,23,24]

Second, we analyzed the morphology of MSC, and observed fibroblast-like cells in BM. Thus the MSCs formed a monolayer of homogenous spindle-shaped cells with a whirlpool-like array. This result is in accordance with the literature. [25]. However, two cell phenotypes were observed among fibroblast-like BM-MSC colonies: thin spindle shaped MSC and star-shaped MSC. These BM-MSC two phenotypes were described by Wolgang et Wagnera and al [25]

To detect the expression of galectins Gal-1, Gal-9, Gal-11 and Gal-13 by MSCs, we used two techniques which are the flow cytometry (FC) and the RT-PCR. We searched the expression of galectins in MSCs cultivated in the three culture mediums: the standard medium: alpha-MEM+ FGF2, the medium (with 5% HPL), the medium (with 10% HPL).

The flow cytometry revealed that all the galectins of interest (Gal-1, Gal-3, Gal-9, and Gal-13) are expressed in BM-MSCs, in above mentioned culture mediums,. It thus becomes clear that expression percents of galacetins revealed by the flow cytometry in hBM-MCSs depend on two parameters which are the composition of the medium and the type of the galectin.

Thus, we observed mainly that the best Galectins' expression was visualized in the standard medium (alpha-MEM /FGF2). Previous studies have shown that the supplementation of fibroblast growth factor 2 (FGF2) in vitro selects for the survival of a large number of of MSCs enriched in pluripotent mesenchymal precursors enhances the growth of MSCs maintains their multilineage differentiation potential during in vitro expansion and prolongs their life span [26]. So, this data and our observation suggest that the increase of galectins in the presence of FGF2 is due to the improvement of the growth the proliferation and differentiation potential of MSCs.

Then, when comparing the means of the galectins' expression by the MSCs we concluded that the best means of protein expression were those of Gal-1; this galectin had the highest means of expression in all culture mediums. Friederike Gieseke and al. confirm our result; thus, they report that Gal-1 is highly expressed in MSCs. [27]

The results of the RT-PCR are in accordance with those of the flow cytometry. Thus, the RT-PCR applied to RNA of the MSCs revealed that galectins (Gal-1, Gal-9, Gal-11, and Gal-13) are expressed by both undifferentiated and differentiated BM-MSCs. So, before the induction of the differentiation process, the MSCs cultivated in the standard medium expressed Gal-1, Gal-9, Gal-11, but not Gal-13, while the MSCs cultivated in the mediums enriched by the HPL expressed only the Gal-1. Nevertheless, after the induction of the MSCs differentiation, first we observed that Gal-1, Gal-9, Gal-11 were no more expressed in the standard culture's medium (alpha MEM+ FGF2) while we detected the expression of Gal-13 in the adipocytes. Second, when examining the expression of galectins by MSCs cultivated in mediums containing the HPL we noticed the presence of Gal-1, Gal-9 with a low expression, Gal-11 in both adipocytes and VMLs and finally Gal-13 in adipocytes.

We noted that MSCs expressed the galectins constitutively, but this expression was modulated differently depending on the type of the galectin the culture's stage and the composition of the culture medium.

These findings are in agreement with previous reports. Thus, Friederike Gieseke and al. [27] claim that m RNAs of Gal-1, Gal-3, -8 and -9 were detected in MSCs by the RT-PCR and when performing a quantitative real time –PCR the m RNAs of Gal-3, -8 and -9 were detected to a lesser extent than Gal-1 which is similar to the results we obtained when performing the flow cytometry.

Najar M and al. [14] confirmed the constitutive expression of galectin-1 m RNA by BM-MSCs. Regarding Gal-3 Ju-Yeon Kim and al. [28] demonstrated that Gal-3 is secreted by UCB-MSC under pathological conditions whereas low level of GAL-3 was detected by Western blot analysis in conditioned medium of naïve UCB-MSC. This report is in contrast with our results since we confirmed the absence of galectins' in both membranes and RNAs of UC-MSCs.

The present study provides new insights concerning the expression of the galectins by the MSCs. According to our observations and the reports mentioned above the galectins are located both in nuclear and membrane levels. This suggests that the galectins are released by MSCs. Because of their their multilineage differentiation potential, immunomodulatory and anti-inflammatory properties, MSCs have become increasingly attractive as a therapeutic approach. Careful characterization of MSC physiology and the effects of culture mediums context on their proliferation differentiation and molecules' secretion will increase the safety and efficiency of MSCs in clinical settings.

Author details

Faouzi Jenhani[1,2*]

Address all correspondence to: faouzi.jenhani@yahoo.tn, tunisiacelltherapy@ymail.com

1 Cell Immunology and Cytometry and Cell Therapy Laboratories, Blood National Center, Tunisia

2 Unit Immunology Research, Faculty of pharmacy, Monastir, Tunisia

References

[1] Tse, W. T. , Pendleton, J. D. , Beyer, W. M. , Egalka, M. C. , Guinan, E. C. Suppression of allogeneic T-cell proliferation by human marrow stromal cells: implications in transplantation. Transplantation,2003, 75, 389–397.

[2] Selmani Z, Naji A, Zidi I, Favier B, Gaiffe E, Obert L, et al. Human leukocyte antigen-G5 secretion by human mesenchymal stem cells is required to suppress T lymphocyte and natural killer function and to induce CD4 1 CD25 high FOXP31 regulatory T cells. Stem Cells 2008;26:212-222.

[3] Le Blanc K, Rasmusson I, Sundberg B, Gotherstrom C, Hassan M, Uzunel M, et al. Treatment of severe acute graft-versus-host disease with third party haploidentical mesenchymal stem cells. Lancet. 2004;363:1439–1441.

[4] Frank MH, Sayegh MH. Immunomodulatory functions of mesenchymal stem cells. Lancet. 2004; 363:1411–2.

[5] Danguy, A. , Camby, I. , Kiss, R. , 2002. Galectins and cancer. Biochim. Biophys. Acta 1572, 285–293.

[6] Yagi H, Soto-Gutierrez A, Parekkadan B, Kitagawa Y, Tompkins RG, Kobayashi N, Yarmush ML: Mesenchymal stem cells: mechanisms of immunomodulation and homing. *Cell Transplant* 2010, 19:667-679.

[7] Pittenger MF, Mackay AM, Beck SC, Jaiswal RK: Multilineage potential of adult human mesenchymal stem cells. *Science* 1999, 284:143-147.

[8] Maria Virginia Tribulatti, Galectin-8 provides costimulatory and proliferative signals to T lymphocytes. Journal of Leukocyte Biology 2009,Volume 86, August.

[9] Rabinovich GA. Galectins: an evolutionarily conserved family of animal lectins with multifunctional properties; a trip from the gene to clinical therapy. Cell Death Differ Aug 1999; 6(8):711–721.

[10] Camby I, Le Mercier M, Lefranc F, Kiss R. Galectin-1: a small protein with major functions. Glycobiology. 2006;16:137R–157R.

[11] D. Baksh, L. Song R. S. Tuan. Adult mesenchymal stem cells : Characterization differentiation and application in cell and gene therapy. J. Cell. Mol. Med. Vol 8, No 3, 2004 pp. 301-316.

[12] Deans RJ, Moseley AB. Mesenchymal stem cells: biology and potential clinical uses. Exp Hematol. 2000;28:875–884.

[13] Camby I, Le Mercier M, Lefranc F, Kiss R. Galectin-1: a small protein with major functions. Glycobiology. 2006;16: 137R–157R.

[14] Najar et al : Mehdi Najar, Gordana Raicevic, Hicham Id Boufker, Basile Stamatopoulos, Ce´cile De Bruyn, Nathalie Meuleman, Dominique Bron, Michel Toungouz, and Laurence Lagneaux. Modulated expression of adhesion molecules and galectin-1: Role during mesenchymal stromal cell immunoregulatory functions. Experimental Hematology 2010;38:922–932.

[15] Capelli C, Domenghini M, Borleri G, Bellavita P, Poma R, Carobbio A, et al. Human platelet lysate allows expansion and clinical grade production of mesenchymal stro-

mal cells from small samples of bone marrow aspirates or marrow filter washouts. Bone Marrow Transplant 2007;40(8):785–91.

[16] Doucet C, Ernou I, Zhang Y, Llense JR, Begot L, Holy X, et al. Platelet lysates promote mesenchymal stem cell expansion: a safety substitute for animal serum in cell-based therapy applications. J Cell Physiol 2005;205(2):228–36.

[17] M Dominici K Le Blanc I Mueller I Slaper-Cortenbach FC Marini DJ Prockop and EM Horwitz. Minimal Criteria for defining multipotent mesenchymal stromal cells. The international society for cellular therapy position statement. Cytotherapy (2006) Vol 8, No. 4, 315-317.

[18] Horwitz EM, Le Blanc K, Dominici M, et al. Clarification of the nomenclature for MSC: The International Society for Cellular Therapy position statement. Cytotherapy. 2005; 7:393–395.

[19] Berger,M. G. ,Veyrat-Masson,R. ,Rapatel,C. ,Descamps, S. , Chassagne,J. ,Boiret-Dupre,N. ,2006. Cellculture medium compositionandtranslationaladultbone marrow-derived stemcellresearch. StemCells24 (12), 2888–2890.

[20] Harrison P, Cramer EM. Platelet alpha-granules. Blood Rev 1993;7(1):52–62.

[21] Kilian O, Flesch I, Wenisch S, Taborski B, Jork A, Schnettler R, et al. Effects of platelet growth factors on human mesenchymal stem cells and human endothelial cells in vitro. Eur J Med Res 2004;9(7):337–44.

[22] Van den Dolder J, Mooren R, Vloon AP, Stoelinga PJ, Jansen JA. Platelet-rich plasma: quantification of growth factor levels and the effect on growth and differentiation of rat bone marrow cells. Tissue Eng 2006;12(11):3067–73.

[23] Capelli C, Domenghini M, Borleri G, Bellavita P, Poma R, Carobbio A, et al. Human platelet lysate allows expansion and clinical grade production of mesenchymal stromal cells from small samples of bone marrow aspirates or marrow filter washouts. Bone Marrow Transplant 2007;40(8):785–91.

[24] Doucet C, Ernou I, Zhang Y, Llense JR, Begot L, Holy X, et al. Platelet lysates promote mesenchymal stem cell expansion: a safety substitute for animal serum in cell-based therapy applications. J Cell Physiol 2005;205(2):228–36.

[25] Comparative characteristics of mesenchymal stem cells from human bone marrow, adipose tissue, and umbilical cord blood. Wolfgang Wagnera, Frederik Weina, Anja Seckingera, Maria Frankhauserb, Ute Wirknerc, Ulf Krausea, Jonathon Blakec, Christian Schwagerc, Volker Ecksteina, Wilhelm Ansorgec, and Anthony D. Hoa. Experimental Hematology 33 (2005) 1402–1416

[26] Bianchi et al : Ex vivo enrichment of mesenchymal cell progenitors by fibroblast growth factor 2. Giordano Bianchi,a,b,c,2 Andrea Banfi,a,b,c,1,2 Maddalena Mastrogiacomo,a,b,c Rosario Notaro,a Lucio Luzzatto,a Ranieri Cancedda,a,b,c and Rodolfo Quartoa,b,c,*. Experimental Cell Research 287 (2003) 98–105.

[27] Friederike Gieseke and al : Friederike Gieseke, Judith Böhringer Rita Bussolari Massimo Dominci Rupert Handgretinger and Ingo Müller. Human multipotent mesenchymal stromal cells employ galectin-1 to inhibit immune effector cells. Jul 19, 2010; doi: 10. 1182/Blood-2010-02-270777.

[28] Ju-Yeon Kim and al: Galectin-3 secreted by human umbilical cord blood-derived mesenchymal stem cells reduces amyloid-b42 neurotoxicity in vitro Ju-Yeon Kim a,b, Dong Hyun Kim a, Dal-Soo Kim a, Ji Hyun Kim a, Sang Young Jeong a, Hong Bae Jeon a, Eun Hui Lee b, Yoon Sun Yang a, Wonil Oh a, Jong Wook Chang a,* FEBS Letters 584 (2010) 3601–3608.

Chapter 4

Canonical HSC Markers and Recent Achievements

Takao Sudo, Takafumi Yokota, Tomohiko Ishibashi,
Michiko Ichii, Yukiko Doi, Kenji Oritani and
Yuzuru Kanakura

Additional information is available at the end of the chapter

http://dx.doi.org/10.5772/54474

1. Introduction

A specific feature of hematopoietic stem cells (HSC) is the potency to supply all types of blood cells throughout life by self-renewal and differentiation. Bone marrow (BM) is actively producing differentiated blood cells with enormous cellular turnover. Under homeostatic state, primitive HSC in adult BM divide only rarely and are located in specialized regulatory environment to avoid exhaustion and DNA damages that are supposed to cumulatively develop hematopoietic disorders such as myelodysplastic syndrome or leukemia. However, those quiescent HSC can be proliferative on demand, particularly on systemic infection or myelo-suppressive treatment. Therefore, elaborate mechanisms regulating the self-renewal and differentiation of BM HSC is indispensable to maintain normal hematopoiesis throughout life. The fluctuating feature of HSC is thought to be associated with their regulatory environment, generally called "HSC niche".

Technical improvement for purifying authentic HSC from heterogeneous cellular populations is necessary to understand the features of those extremely rare and precious cells and promote their therapeutic application. Many studies have attempted to identify their specific markers, and now flow cytometry- based strategies have made it possible to sort HSC with high purity in mice. However, the source and the stage of HSC change along ontogeny, which consequently influence not only their functional abilities but also their surface immunophenotypes. In the light of fluctuating nature of HSC, it should be very important to understand their phenotype specific to reconstitution activity of the immune system after myelo-suppressive events.

Hematopoietic cells and endothelial cells are both generated from mesodermal precursor cells in ontogeny [1]. Thereafter, HSC pool is formed in several anatomical sites such as aorta-gonad-mesonephros (AGM) region, placenta, fetal liver, and BM. At the early stage of ontogeny, HSC frequently undergo symmetrical and/or asymmetrical division to form entire hematopoietic system compared to adult HSC. Those early HSC and endothelial cells bear some common surface antigens, of which expression levels on HSC decline along aging. Interestingly, some of the endothelial-related surface molecules revive on HSC after BM injury, when the cells actively divide to regenerate BM cells.

We recently reported endothelial cell-selective adhesion molecule (ESAM) as a new marker for HSC [2]. Interestingly, ESAM levels on HSC clearly mirror the shift of HSC between quiescence and activation, and the up-regulation amplitude is prominent in comparison to other HSC-related antigens [3]. Furthermore, we found that ESAM is functionally indispensable for HSC to re-establish homeostatic hematopoiesis [3]. In this chapter, we review a wealth of information about traditional HSC markers, and introduce our recent findings.

2. Development of the strategy for purifying HSC from murine BM

HSC are defined by their capacity for both self-renewal and differentiation into all the blood cell types. In 1988, Spangrude *et al.* reported lineage (Lin; Mac-1, Gr-1, B220, CD4, and CD8)$^-$ Thy-1^{Lo} Sca-1$^+$ cells in mouse BM as a multipotent HSC population. When these cells were transplanted into lethally irradiated mice, only thirty cells were sufficient to save 50% of the recipient mice and reconstitute B, T, and myeloid cells [4]. In 1991, Ogawa *et al.* reported that half of the c-kit$^+$ BM cells do not express Lin (Mac-1, Gr-1, Ter119, and B220) markers, and c-kit$^-$ population do not include hematopoietic progenitor cells [5]. From then on, Lin$^-$ Sca-1$^+$ c-kit$^+$ (LSK) cells has been used as the population in which HSC are highly concentrated [6,7]. HSC can be functionally classified as either long-term (LT-HSC) or short-term (ST-HSC) according to their capacity to give rise to life-long or transient hematopoiesis. Osawa *et al.* showed that CD34$^+$ LSK cells are capable of only short-term multilineage differentiation. In contrast, CD34$^{-/Lo}$ LSK cells have long-term multilineage reconstitution capacity. They also showed that CD34$^{-/Lo}$ LSK cells can differentiate into CD34$^+$ LSK cells [7]. LSK fraction also can be divided into two populations by expression level of Flk-2. While LT-HSC are enriched in the Flk-2$^-$ LSK fraction, the Flk-2$^+$ LSK cells are mainly ST-HSC [8].

While the techniques of purifying HSC by use of surface markers had been promoted, Goodell *et al.* reported the method for purifying HSC without use of surface markers. Hoechst33342 is a fluorescent dye which binds to DNA of live cells. When Hoechst fluorescence on whole BM was examined simultaneously at two emission wavelength (red and blue), one population of cells with increased ability to efflux Hoechst dye was observed. Goodell *et al.* named it "side population (SP)", and showed that a majority of HSC were enriched in the SP by competitive repopulating experiments [9]. Subsequently, Matsuzaki *et al.* described a method of further purifying HSC by combining staining with antibodies to surface molecules with the Hoechst dye efflux. They showed the fraction of cells with the strongest dye efflux activity (termed as "Tip"-SP) has the highest marrow-repopulating activity. While 20% of "Tip"-SP cells are primitive hematopoietic cells, more than 90% of "Tip"-SP

CD34⁻ KSL cells, which are extremely rare, representing only 0.001-0.01% of BM mononuclear cells, are almost pure primitive hematopoietic cells that have long-term multilineage repopulating potency [10]. More recently, the endothelial protein C receptor CD201 was found as a new endothelial-related HSC marker which marks approximately 70% of the SP cells. The marker seems to be useful to purify LT-HSC among the SP cells because only the $CD201^{+}$ subpopulation exhibited repopulating ability [11].

In recent years, Kiel *et al.* demonstrated that a simple combination of SLAM family markers (CD150, CD244 and CD48) could enrich primitive murine HSC. That is, one out of every 4.8 (21%) of $CD150^{+}$ $CD48^{-}$ cells from young adult murine BM gave long-term multilineage reconstitution [12]. Furthermore, they observed that one out of every 2.1 (47%) of $CD150^{+}$ $CD48^{-}$ LSK cells had long-term multilineage reconstituting potential. Approximately 15~20% of $CD150^{+}$ $CD48^{-}$ LSK or $CD34^{-}$ $CD150^{+}$ $CD48^{-}$ LSK cells, which divide only 5-6 times during the mouse life span, have more long-term repopulating potential than other cells [13,14]. We can now purify dormant LT-HSC from murine BM using the SLAM family markers in combination with LSK gating. The information regarding murine HSC markers is summarized in Table 1.

Markers	References
Lin^{-} $Thy\text{-}1^{Lo}$ $Sca\text{-}1^{+}$	Spangrude *et al. Science* (1988) [4]
$CD34^{-/Lo}$ LSK	Osawa *et al. Science* (1996) [7]
Side population (SP)	Goodell *et al. J Exp Med* (1996) [9]
Lin^{-} Rho^{-} SP	Uchida *et al. Exp Hematol* (2003) [15]
Tip-SP LSK	Matsuzaki *et al. Immunity* (2004) [10]
$CD48^{-}$ $CD150^{+}$ $CD41^{-}$ LSK	Kiel *et al. Cell* (2005) [12]
BrdU and histone 2B-retaining $CD48^{-}$ $CD150^{+}$ $CD34^{-}$ LSK	Wilson *et al. Cell* (2008) [14]
Histone 2B-retaining $CD48^{-}$ $CD150^{+}$ LSK	Foudi *et al. Nat Biotechnol* (2009) [13]

Lin, lineage; Rho, Rhodamine-123; LSK, Lin^{-} $Sca\text{-}1^{+}$ $c\text{-}kit^{+}$

Table 1. Markers for adult murine hematopoietic stem cells.

3. HSC markers during developmental stages

HSC markers during developmental stages are not identical to those of adult HSC. In the embryo, functional HSC that can reconstitute hematopoiesis in adult recipients are firstly found in the aorta-gonad-mesonephros (AGM) region at approximately embryonic day 10 (E10) [16-18]. Many reports have demonstrated that those earliest authentic HSC bud from endothelial-related cells, which involve the concept of "hemangioblast" or "hemogenic endothelium" [19-23]. In fact, emerging HSC and endothelial cells share various surface markers such as CD34 and VE-cadherin that do not mark adult murine LT-HSC [24-26]. On the contrary, the emerging HSC do not express either Sca-1 or CD45, a pan-hematopoietic marker [19,27]. Interestingly, those developing HSC express CD41/Integrin-α_v, a marker for megakaryocytes [28].

Although HSC do not emerge in the fetal liver de novo, the organ is the main site of HSC expansion before birth. Circulating HSC seed in the fetal liver, where they robustly expand and differentiate. Indeed, numbers of HSC increase ~40-fold in the fetal liver between E12 and E16 [29]. Unlike the emerging HSC in the AGM region, HSC in fetal liver express CD45 and Sca-1. Morrison *et al.* showed that HSC are highly enriched in Thy1^{Lo} Sca-1$^+$ Lin$^-$ Mac1$^+$ fraction of fetal liver cells [30]. His group later demonstrated that the SLAM family markers (CD150$^+$ CD48$^-$) are also useful to enrich for HSC in E14.5 fetal liver just as in adult BM by the fact that 37% of CD150$^+$ CD48$^-$ Sca-1$^+$ Lin$^-$ Mac1$^+$ fetal liver cells had long-term reconstituting capacity [31]. Although the expression levels of AA4.1 and VE-cadherin are very high at the early stage of fetal hematopoiesis, they become gradually down-regulated after E12-E13 [32,33]. Interestingly, the phenotype of HSC in fetal liver rapidly changes after E16, when their number is reaching to a plateau level [30]. Recently, our group reported ESAM as a novel HSC marker in fetal liver (see below). HSC markers during mouse ontogeny are summarized in Table 2.

Fetal age	Location	Markers	References
E8.5-E10.5	AGM region	CD41$^+$ CD34$^+$ CD45$^-$ VE-cadherin$^+$ Sca-1$^-$ AA4.1$^+$ ESAM$^+$	Petrenko *et al. Immunity* (1999) [33], Hsu *et al. Blood* (2000) [34], Baumann *et al. Blood* (2004) [24], Fraser *et al. Exp Hematol* (2002) [25], Ogawa *Exp Hematol* (2002) [26], de Brujin *et al. Immunity* (2002) [19], Mikkola *et al. Blood* (2003) [28], Matsubara *et al. J Exp Med* (2005) [27], Kim *et al. Blood* (2005) [32], Kim *et al. Blood* (2006) [31], Mansson *et al. Immunity* (2007) [35], Yokota *et al.* Blood (2009) [2]
E11.5-E16.5	Fetal liver	CD41$^-$ CD34$^+$ CD45$^+$ CD31$^+$ Sca-1$^+$ Mac1$^+$ Tie-2$^+$ Flt3$^-$ c-kit$^+$ AA4.1$^+$ VE-cadherin$^{+/-}$ CD150$^+$ CD48$^-$ ESAM$^+$	

Table 2. Markers for hematopoietic stem cells during mouse ontogeny.

Adult and fetal HSC are not the same with regard to not only surface phenotypes but also cell-cycle status. Recent studies have shown that the long-term reconstituting activity of adult BM is sustained mostly in very quiescent HSC [13,14]. However, cycling HSC from the fetal liver give rise to higher levels of reconstitution than HSC obtained from adult BM [30,36]. The microenvironments, known as "HSC niches", are believed to influence cell-cycle status of HSC, and adult HSC niches in BM seem to be different from HSC niches in the fetal liver [12,37-40]. More precise analyses of hematopoietic environment in the embryo should give us valuable information regarding what are the imperative conditions for HSC expansion and how the alteration of surface molecules on HSC is functionally involved in that process. Furthermore, such cell surface antigens that mirror the HSC state are invaluable for understanding the relationship between HSC and their niches.

http://dx.doi.org/10.5772/54474

4. Niche signals regulating HSC pool

We think it seems meaningful to deal with the "HSC niche" briefly here, although another chapter in this book provides more detailed information about its function. Molecular cross-talk between HSC and their niches has been considered to be important to provide signals for self-renewing division that maintain HSC pool. Although precise mechanisms regulating HSC status still remain unknown, there are accumulating evidences to involve several specific cells, or cytokines and chemokines secreted from stromal cells in this process.

In 1994, human osteoblasts were shown to maintained hematopoiesis by constitutively producing G-CSF in vitro [41]. In the first decade of the 21st century, a notion that connects osteoblasts with the HSC niche rapidly developed. Parathyroid hormone (PTH), which is a main regulator of calcium homeostasis, was reported to increase in the number of both osteoblasts and HSC, suggesting osteoblasts as the candidate for HSC niche [42]. In addition, it was also reported that BrdU label retaining cells (LRC) were attached to spindle-shaped N-cadherin^{+} osteoblasts (SNO) cells, and that bone morphogenetic protein (BMP) signalling controlled the number of HSC by regulating SNO cells [37].

On the other hand, Kiel *et al.* reported that many CD150^{+} CD48^{-} CD41^{-} Lin^{-} LT-HSC were in contact with sinusoidal endothelial cells in spleen or BM, suggesting that endothelial cells are also essential components of the HSC niche. With regard to cytokine-chemokine sinalings, the CXC chemokine ligand 12 (CXCL12) -CXC chemokine receptor 4 (CXCR4) pathway was found to be important. In vitro, HSC expressing CXCR4 migrate in response to CXCL12 which is the ligand for CXCR4 [43]. Nagasawa's laboratory reported that a majority of CD150^{+} CD48^{-} CD41^{-} HSC were in contact with CXCL12-abundant reticular (CAR) cells, and that the numbers of HSC in CAR cell-depleted mice were reduced in comparison with control mice. These data are supportive of the idea that CXCL12-CXCR4 pathway is essential for HSC pool [39,44]. Recently, Yamazaki *et al.* reported TGF- β as a candidate niche signal in the control of HSC hibernation [45]. The same group advocated that glial cells, regulating activation of TGF-β signal, might be a component of the HSC niche in adult BM and maintain HSC hibernation [46].

5. Differences between murine and human HSC markers

A critical issue that has been an obstacle in applying the information of murine HSC to human is the lack of common HSC markers between the two species. Researchers described above have made great efforts to purify authentic HSC from murine hematopoietic organs. Owing to those achievements, we can now sort LT-HSC with very high purity from the murine BM. However, human HSC cannot be purified with the same markers. Human HSC do not express Sca-1 or CD150 that are the established HSC markers in mice. In addition, the long-term HSC of human BM are enriched in CD34^{+} CD38^{-} population, while murine BM HSC are CD34^{-} CD38^{+} [26,47,48].

Early studies in the 1980s proved by using monoclonal antibody technique that CD34$^+$ population of human BM includes immature hematopoietic progenitors [49-51]. Berenson *et al.* showed that autologous CD34$^+$ cells enriched from baboon BM were able to reconstitute normal hematopoiesis after lethal irradiation. Animals transplanted with CD34$^-$ cells, however, did not recover sufficient hematopoiesis [52]. Afterwards, over the past two decades, CD34-positive has been used as a reliable marker for human HSC or hematopoietic progenitor cells (HPC). Indeed, transplantation of CD34$^+$ cells obtained from donor BM, peripheral blood, or cord blood (CB) can provide long-term and multilineage hematopoietic reconstitution in recipients.

As CD34 marks human HSC as well as more differentiated progenitor cells, researchers have sought additional markers to further enrich CD34$^+$ population for LT-HSC. Baum *et al.* reported that CD90/Thy-1$^+$ population in Lin$^-$ CD34$^+$ cells contained pluripotent hematopoietic progenitors [53]. Recently a series of studies of John Dick's laboratory have successfully improved the techniques to more purify human HSC. His group reported that human HSC activity was restricted to CD49f$^+$ fraction, and that single Lin$^-$ CD34$^+$ CD38$^-$ CD45RA$^-$ Thy-1$^+$ Rhodamin123^{Lo} CD49f$^+$ cells in CB cells accomplished multilineage engraftment in immune-deficient mice [54].

While LT-HSC can be enriched mainly in the CD34$^+$ population, the possibility that CD34$^-$ cells also contain LT-HSC has been reported. Bhatia *et al.* showed human CD34$^-$ population in Lin$^-$ cells of BM and CB also contained LT-HSC [55]. It should be important to compare the features of primate CD34$^-$ HSC with those of murine CD34$^-$ LSK cells. In addition, a new positive marker for human HSC could resolve the relationship between the CD34$^+$ and CD34$^-$ HSC. Markers for human HSC are summarized in Table 3.

Markers	References
CD34$^+$	Berenson *et al. J Clin Invest* (1988) [52]
CD34$^+$ CD38$^-$	Terstappen *et al. Blood* (1991) [48]
CD34$^+$ Lin$^-$ Thy-1$^+$	Baum *et al. Proc Natl Acad Sci USA* (1992) [53]
Lin$^-$ CD34$^-$ CD38$^-$	Bhatia *et al. Nat Med* (1998) [55]
CD34$^+$ CD38$^-$ Lin$^-$ Rho^{Lo}	McKenzie *et al. Blood* (2007) [47]
Lin$^-$ CD34$^+$ CD38$^-$ CD45RA$^-$ Thy-1$^+$ Rho^{Lo} CD49f$^+$	Notta *et al. Science* (2011) [54]

Table 3. Markers for human hematopoietic stem cells.

6. Differences between quiescent and activated HSC markers

After mice are treated with cytotoxic agents or irradiation, most of cell-cycling hematopoietic cells are killed and dormant primitive HSC start to proliferate. The patterns of surface molecules expressed on activated HSC change from those under steady-state condition. While activated HSC increase the expression level of Sca-1, CD150, Tie2, Endoglin, Mac-1, and CD34, they clearly decrease that of c-kit and N-cadherin [26,56,57]. Some endothelial-related antigens, which mark actively dividing fetal HSC but do not mark quiescent adult HSC, are up-regulated again on the activated HSC after BM injury. The characteristics of those activated HSC are reminiscent of fetal HSC. Since no obvious phenotypes have been documented regarding CD34 or CD150-deficient mice, how the up-regulation of those molecules contributes to the functions and/or characteristics of activated HSC remains unknown [12,58]. Tie2 and Endoglin, which are the receptors for angiopoietin and TGF, respectively, might transduce important signals to regulate dividing speed of HSC. If we could accurately monitor the fluctuation of HSC status with a set of surface markers, that should yield significant insight regarding HSC biology and HSC applications for clinical purposes. As a very recent achievement, our group has demonstrated that ESAM is a useful marker for activated HSC.

7. An endothelial-related antigen ESAM as a new novel HSC marker

We previously reported sorting strategy of HSC and early lymphoid progenitors (ELP) from Rag1/GFP knockin mice [59,60]. We searched for genes whose expression levels are significantly different between $Rag1^{-}$ $c\text{-}kit^{Hi}$ $Sca\text{-}1^{+}$ HSC and $Rag1^{Lo}$ $c\text{-}kit^{Hi}$ $Sca\text{-}1^{+}$ ELP by analyzing micro-array data. Among the HSC related genes ESAM drew our attention because its transcripts were conspicuous in the HSC fraction whereas the expression was drastically down-regulated in the ELP fraction. ESAM molecule is an immunoglobulin superfamily protein that is exposed on cell surface and originally identified as an endothelial cell-specific protein [61,62]. We found the $ESAM^{Hi}$ population of $Rag1^{-}$ $c\text{-}kit^{Hi}$ $Sca\text{-}1^{+}$ fraction of E14.5 fetal liver was highly enriched for LT-HSC compared with $ESAM^{-/Lo}$ subset. Among $Rag1/GFP^{-}$ $Tie2^{Hi}$ E10.5 AGM cells, only $ESAM^{+}$ cells could effectively produce both $CD19^{+}$ lymphoid cells and $Mac1^{+}$ myeloid cells [2].

ESAM is also expressed on adult murine HSC-enriched fraction in BM. Ooi *et al.* reported that $ESAM^{+}$ $Sca\text{-}1^{+}$ Lin^{-} BM cells could more effectively enrich for LT-HSC than the conventional HSC-enriched LSK cells, and that ESAM expression on HSC was conserved among various mouse strains [63]. ESAM levels on HSC are variable according to developing stages or advancing age. Interestingly, the intensity of ESAM expression on HSC gradually increased with age after reaching adulthood [2]. Based on these observations, ESAM can be a novel murine HSC marker throughout life including developmental stages.

The usefulness of ESAM as a HSC marker has been further enhanced by the findings that its expression in human HSC is also detected. Ooi *et al.* reported that robust levels of ESAM transcripts were detected in Lin^{-} $CD34^{+}$ $CD38^{-}$ $CD90^{+}$ human HSC, while the levels of ESAM transcripts in unfractionated CB cells were very low [63]. We have confirmed that ESAM expression is clearly detectable on human CB $CD34^{+}$ cells by using its specific antibody and flow cytometry [64]. In addition, our group has also observed that the marker is effective as well for adult human HSC in both BM and mobilized peripheral blood. (Ishibashi *et al.* manuscript in preparation).

8. ESAM monitors HSC status between quiescence and self-renewal

As mentioned above, the expression pattern of surface antigens on activated HSC after BM injury substantially differs from that on quiescent HSC. Administration of an anti-cancer drug 5-FU causes apoptosis of dividing hematopoietic progenitors, while the treatment retains quiescent LT-HSC and induces their proliferation afterward. We have observed that remarkable increase of ESAM expression levels transiently occurs on BM HSC after a 5-FU treatment. Furthermore, we have proved that the long-term hematopoietic reconstituting activity is almost exclusive to LSK cells bearing up-regulated ESAM expression [3].

Although expression levels of CD34, Tie2, and Endoglin on LSK show modest increases after 5-FU injection, up-regulation of ESAM is remarkable (Figure 1). Why does ESAM need to revive so vividly on HSC after BM injury? One possible reason is that HSC might directly receive necessary signals which regulate self-renewal or differentiation via interaction with ESAM. Another possibility is that high amounts of ESAM might change the polarity or mobility of HSC, which consequently facilitate them to settle in adequate supporting niches (Figure 2). The latter assumption is likely because Wegmann and colleagues reported that ESAM deficiency causes insufficient Rho signalling in endothelial cells, which regulates the stabilization of endothelial tight junctions [65]. Rho is also expressed in hematopoietic progenitors and involved in their polarity and mobility [66]. It is noteworthy that more than 80% of ESAM^{Hi} HSC were located around perivascular areas in 5-FU-treated BM[3].

In any case, ESAM is likely to play an indispensable role during the recovery from BM injury. Because, while ESAM deficient mice do not show significant hematopoietic defects in homeostatic stage, the mice fall into severe and prolonged pancytopenia after myelo-suppressive treatment. In particular, they suffer from severe anemia and frequently die before hematopoietic recovery. Our findings indicate that ESAM not only marks activated HSC but also functionally supports their proliferation and differentiation.

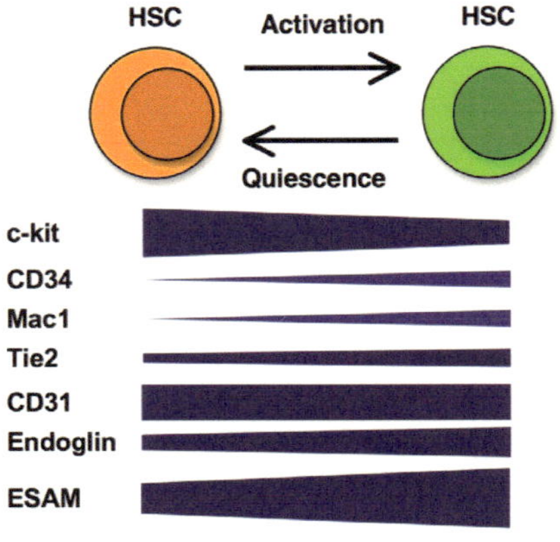

Figure 1. Overview of cell surface expression levels on quiescent steady-state HSC and activated HSC.

http://dx.doi.org/10.5772/54474

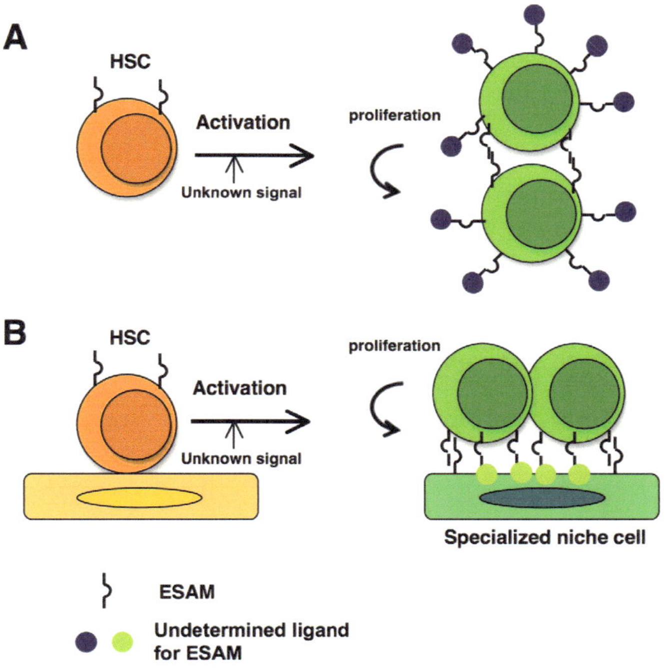

Figure 2. Tentative models of ESAM function. (A) In this model, activated HSC directly receive necessary signals which regulate self-renewal or differentiation via interaction with ESAM. (B) In this model, HSC change their polarity or mobility, and consequently, they can move to appropriate niches. ESAM may function as an adhesion factor between HSC and their niches.

9. Concluding remarks

In this chapter, we summarized achievements for identification of murine and human HSC, and introduced endothelial-related antigen ESAM as a useful HSC marker. While we can now purify murine LT-HSC with high efficiency, characterization of human HSC is less well understood because of insufficient information about surface antigens. Over two decades CD34-positive has been believed to be a reliable marker for human HSC/HPC. Although there are now accumulating evidences regarding surface markers to further enrich human LT-HSC in the $CD34^+$ fraction, more information about human HSC-related antigens should be useful to improve strategies of HSC application to the clinical medicine. Although ESAM was originally identified with an endothelial specific molecule, we have demonstrated that it is a positive marker for both murine and human HSC. Because ESAM seems to play an essential role for hematopoietic recovery after BM injury, it would be significant to elucidate downstream signals of ESAM, and the possibility of ESAM as niche components. In addition, we now know that the up-regulation of ESAM is observed on cultured murine embryonic stem (ES) cells cultured in the OP9 system which recapitulate very primitive stages of hematopoietic development [67] (Doi et al. manuscript in preparation). ESAM might have some roles in embryonic hematopoiesis at very early stages. As an on-going study, we are now investigating whether ESAM can be a useful biomarker for inducing hematopoietic cells from ES or induced pluripotent stem cells.

Author details

Takao Sudo, Takafumi Yokota*, Tomohiko Ishibashi, Michiko Ichii, Yukiko Doi, Kenji Oritani and Yuzuru Kanakura

*Address all correspondence to: yokotat@bldon.med.osaka-u.ac.jp

Department of Hematology and Oncology, Osaka University Graduate School of Medicine, Suita, Japan

References

[1] Choi K, Kennedy M, Kazarov A, et al. A common precursor for hematopoietic and endothelial cells. Development. 1998;125:725-732.

[2] Yokota T, Oritani K, Butz S, et al. The endothelial antigen ESAM marks primitive hematopoietic progenitors throughout life in mice. Blood. 2009;113:2914-2923.

[3] Sudo T, Yokota T, Oritani K, et al. The Endothelial Antigen ESAM Monitors Hematopoietic Stem Cell Status between Quiescence and Self-Renewal. J Immunol. 2012;189:200-210.

[4] Spangrude GJ, Heimfeld S, Weissman IL. Purification and characterization of mouse hematopoietic stem cells. Science. 1988;241:58-62.

[5] Ogawa M, Matsuzaki Y, Nishikawa S, et al. Expression and function of c-kit in hemopoietic progenitor cells. J Exp Med. 1991;174:63-71.

[6] Okada S, Nakauchi H, Nagayoshi K, et al. In vivo and in vitro stem cell function of c-kit- and Sca-1-positive murine hematopoietic cells. Blood. 1992;80:3044-3050.

[7] Osawa M, Nakamura K, Nishi N, et al. In vivo self-renewal of c-Kit+ Sca-1+ Lin(low/-) hemopoietic stem cells. J Immunol. 1996;156:3207-3214.

[8] Christensen JL, Weissman IL. Flk-2 is a marker in hematopoietic stem cell differentiation: a simple method to isolate long-term stem cells. Proc Natl Acad Sci U S A. 2001;98:14541-14546.

[9] Goodell MA, Brose K, Paradis G, et al. Isolation and functional properties of murine hematopoietic stem cells that are replicating in vivo. J Exp Med. 1996;183:1797-1806.

[10] Matsuzaki Y, Kinjo K, Mulligan RC, et al. Unexpectedly efficient homing capacity of purified murine hematopoietic stem cells. Immunity. 2004;20:87-93.

[11] Balazs AB, Fabian AJ, Esmon CT, et al. Endothelial protein C receptor (CD201) explicitly identifies hematopoietic stem cells in murine bone marrow. Blood. 2006;107:2317-2321.

[12] Kiel MJ, Yilmaz OH, Iwashita T, et al. SLAM family receptors distinguish hematopoietic stem and progenitor cells and reveal endothelial niches for stem cells. Cell. 2005;121:1109-1121.

[13] Foudi A, Hochedlinger K, Van Buren D, et al. Analysis of histone 2B-GFP retention reveals slowly cycling hematopoietic stem cells. Nat Biotechnol. 2009;27:84-90.

[14] Wilson A, Laurenti E, Oser G, et al. Hematopoietic stem cells reversibly switch from dormancy to self-renewal during homeostasis and repair. Cell. 2008;135:1118-1129.

[15] Uchida N, Dykstra B, Lyons KJ, et al. Different in vivo repopulating activities of purified hematopoietic stem cells before and after being stimulated to divide in vitro with the same kinetics. Exp Hematol. 2003;31:1338-1347.

[16] Cumano A, Ferraz JC, Klaine M, et al. Intraembryonic, but not yolk sac hematopoietic precursors, isolated before circulation, provide long-term multilineage reconstitution. Immunity. 2001;15:477-485.

[17] Medvinsky A, Dzierzak E. Definitive hematopoiesis is autonomously initiated by the AGM region. Cell. 1996;86:897-906.

[18] Muller AM, Medvinsky A, Strouboulis J, et al. Development of hematopoietic stem cell activity in the mouse embryo. Immunity. 1994;1:291-301.

[19] de Bruijn MF, Ma X, Robin C, et al. Hematopoietic stem cells localize to the endothelial cell layer in the midgestation mouse aorta. Immunity. 2002;16:673-683.

[20] de Bruijn MF, Speck NA, Peeters MC, et al. Definitive hematopoietic stem cells first develop within the major arterial regions of the mouse embryo. EMBO J. 2000;19:2465-2474.

[21] Godin I, Garcia-Porrero JA, Dieterlen-Lievre F, et al. Stem cell emergence and hemopoietic activity are incompatible in mouse intraembryonic sites. J Exp Med. 1999;190:43-52.

[22] Tavian M, Coulombel L, Luton D, et al. Aorta-associated CD34+ hematopoietic cells in the early human embryo. Blood. 1996;87:67-72.

[23] Tavian M, Hallais MF, Peault B. Emergence of intraembryonic hematopoietic precursors in the pre-liver human embryo. Development. 1999;126:793-803.

[24] Baumann CI, Bailey AS, Li W, et al. PECAM-1 is expressed on hematopoietic stem cells throughout ontogeny and identifies a population of erythroid progenitors. Blood. 2004;104:1010-1016.

[25] Fraser ST, Ogawa M, Yu RT, et al. Definitive hematopoietic commitment within the embryonic vascular endothelial-cadherin(+) population. Exp Hematol. 2002;30:1070-1078.

[26] Ogawa M. Changing phenotypes of hematopoietic stem cells. Exp Hematol. 2002;30:3-6.

[27] Matsubara A, Iwama A, Yamazaki S, et al. Endomucin, a CD34-like sialomucin, marks hematopoietic stem cells throughout development. J Exp Med. 2005;202:1483-1492.

[28] Mikkola HK, Fujiwara Y, Schlaeger TM, et al. Expression of CD41 marks the initiation of definitive hematopoiesis in the mouse embryo. Blood. 2003;101:508-516.

[29] Ema H, Nakauchi H. Expansion of hematopoietic stem cells in the developing liver of a mouse embryo. Blood. 2000;95:2284-2288.

[30] Morrison SJ, Hemmati HD, Wandycz AM, et al. The purification and characterization of fetal liver hematopoietic stem cells. Proc Natl Acad Sci U S A. 1995;92:10302-10306.

[31] Kim I, He S, Yilmaz OH, et al. Enhanced purification of fetal liver hematopoietic stem cells using SLAM family receptors. Blood. 2006;108:737-744.

[32] Kim I, Yilmaz OH, Morrison SJ. CD144 (VE-cadherin) is transiently expressed by fetal liver hematopoietic stem cells. Blood. 2005;106:903-905.

[33] Petrenko O, Beavis A, Klaine M, et al. The molecular characterization of the fetal stem cell marker AA4. Immunity. 1999;10:691-700.

[34] Hsu HC, Ema H, Osawa M, et al. Hematopoietic stem cells express Tie-2 receptor in the murine fetal liver. Blood. 2000;96:3757-3762.

[35] Mansson R, Hultquist A, Luc S, et al. Molecular evidence for hierarchical transcriptional lineage priming in fetal and adult stem cells and multipotent progenitors. Immunity. 2007;26:407-419.

[36] Harrison DE, Zhong RK, Jordan CT, et al. Relative to adult marrow, fetal liver repopulates nearly five times more effectively long-term than short-term. Exp Hematol. 1997;25:293-297.

[37] Zhang J, Niu C, Ye L, et al. Identification of the haematopoietic stem cell niche and control of the niche size. Nature. 2003;425:836-841.

[38] Wilson A, Trumpp A. Bone-marrow haematopoietic-stem-cell niches. Nat Rev Immunol. 2006;6:93-106.

[39] Sugiyama T, Kohara H, Noda M, et al. Maintenance of the hematopoietic stem cell pool by CXCL12-CXCR4 chemokine signaling in bone marrow stromal cell niches. Immunity. 2006;25:977-988.

[40] Iwasaki H, Arai F, Kubota Y, et al. Endothelial protein C receptor-expressing hematopoietic stem cells reside in the perisinusoidal niche in fetal liver. Blood. 2010;116:544-553.

[41] Taichman RS, Emerson SG. Human osteoblasts support hematopoiesis through the production of granulocyte colony-stimulating factor. J Exp Med. 1994;179:1677-1682.

[42] Calvi LM, Adams GB, Weibrecht KW, et al. Osteoblastic cells regulate the haematopoietic stem cell niche. Nature. 2003;425:841-846.

[43] Wright DE, Bowman EP, Wagers AJ, et al. Hematopoietic stem cells are uniquely selective in their migratory response to chemokines. J Exp Med. 2002;195:1145-1154.

[44] Omatsu Y, Sugiyama T, Kohara H, et al. The essential functions of adipo-osteogenic progenitors as the hematopoietic stem and progenitor cell niche. Immunity. 2010;33:387-399.

[45] Yamazaki S, Iwama A, Takayanagi S, et al. TGF-beta as a candidate bone marrow niche signal to induce hematopoietic stem cell hibernation. Blood. 2009;113:1250-1256.

[46] Yamazaki S, Ema H, Karlsson G, et al. Nonmyelinating Schwann cells maintain hematopoietic stem cell hibernation in the bone marrow niche. Cell. 2011;147:1146-1158.

[47] McKenzie JL, Takenaka K, Gan OI, et al. Low rhodamine 123 retention identifies long-term human hematopoietic stem cells within the Lin-CD34+CD38- population. Blood. 2007;109:543-545.

[48] Terstappen LW, Huang S, Safford M, et al. Sequential generations of hematopoietic colonies derived from single nonlineage-committed CD34+CD38- progenitor cells. Blood. 1991;77:1218-1227.

[49] Andrews RG, Singer JW, Bernstein ID. Monoclonal antibody 12-8 recognizes a 115-kd molecule present on both unipotent and multipotent hematopoietic colony-forming cells and their precursors. Blood. 1986;67:842-845.

[50] Civin CI, Strauss LC, Brovall C, et al. Antigenic analysis of hematopoiesis. III. A hematopoietic progenitor cell surface antigen defined by a monoclonal antibody raised against KG-1a cells. J Immunol. 1984;133:157-165.

[51] Tindle RW, Nichols RA, Chan L, et al. A novel monoclonal antibody BI-3C5 recognises myeloblasts and non-B non-T lymphoblasts in acute leukaemias and CGL blast crises, and reacts with immature cells in normal bone marrow. Leuk Res. 1985;9:1-9.

[52] Berenson RJ, Andrews RG, Bensinger WI, et al. Antigen CD34+ marrow cells engraft lethally irradiated baboons. J Clin Invest. 1988;81:951-955.

[53] Baum CM, Weissman IL, Tsukamoto AS, et al. Isolation of a candidate human hematopoietic stem-cell population. Proc Natl Acad Sci U S A. 1992;89:2804-2808.

[54] Notta F, Doulatov S, Laurenti E, et al. Isolation of single human hematopoietic stem cells capable of long-term multilineage engraftment. Science. 2011;333:218-221.

[55] Bhatia M, Bonnet D, Murdoch B, et al. A newly discovered class of human hematopoietic cells with SCID-repopulating activity. Nat Med. 1998;4:1038-1045.

[56] Haug JS, He XC, Grindley JC, et al. N-cadherin expression level distinguishes reserved versus primed states of hematopoietic stem cells. Cell Stem Cell. 2008;2:367-379.

[57] Randall TD, Weissman IL. Phenotypic and functional changes induced at the clonal level in hematopoietic stem cells after 5-fluorouracil treatment. Blood. 1997;89:3596-3606.

[58] Cheng J, Baumhueter S, Cacalano G, et al. Hematopoietic defects in mice lacking the sialomucin CD34. Blood. 1996;87:479-490.

[59] Igarashi H, Gregory SC, Yokota T, et al. Transcription from the RAG1 locus marks the earliest lymphocyte progenitors in bone marrow. Immunity. 2002;17:117-130.

[60] Yokota T, Kouro T, Hirose J, et al. Unique properties of fetal lymphoid progenitors identified according to RAG1 gene expression. Immunity. 2003;19:365-375.

[61] Hirata K, Ishida T, Penta K, et al. Cloning of an immunoglobulin family adhesion molecule selectively expressed by endothelial cells. J Biol Chem. 2001;276:16223-16231.

[62] Nasdala I, Wolburg-Buchholz K, Wolburg H, et al. A transmembrane tight junction protein selectively expressed on endothelial cells and platelets. J Biol Chem. 2002;277:16294-16303.

[63] Ooi AG, Karsunky H, Majeti R, et al. The adhesion molecule esam1 is a novel hematopoietic stem cell marker. Stem Cells. 2009;27:653-661.

[64] Yokota T, Oritani K, Butz S, et al. Markers for Hematopoietic Stem Cells: Histories and Recent Achievements. Advances in Hematopoietic Stem Cell Research: InTech; 2012:77-88.

[65] Wegmann F, Petri B, Khandoga AG, et al. ESAM supports neutrophil extravasation, activation of Rho, and VEGF-induced vascular permeability. J Exp Med. 2006;203:1671-1677.

[66] Fonseca AV, Freund D, Bornhauser M, et al. Polarization and migration of hematopoietic stem and progenitor cells rely on the RhoA/ROCK I pathway and an active reorganization of the microtubule network. J Biol Chem. 2010;285:31661-31671.

[67] Nakano T, Kodama H, Honjo T. Generation of lymphohematopoietic cells from embryonic stem cells in culture. Science. 1994;265:1098-1101.

Chapter 5

Hematopoietic Stem Cells and Response to Interferon

Atsuko Masumi

Additional information is available at the end of the chapter

http://dx.doi.org/10.5772/54689

1. Introduction

Homeostasis in the bone marrow is dependent on the ability of hematopoietic stem cells (HSCs) to self-renew faithfully, differentiate into various lineages of the hematopoietic system, and form blood cells of several types (Figure 1) [1,2]. Under homeostatic conditions, HSCs are thought to be quiescent, and they are referred to as long-term reconstituting HSCs (LT-HSC) or dormant HSCs (dHSCs) [3,4]. Blood and immune cells are produced by the more differentiated short-term reconstituting HSCs (ST-HSCs) or multipotent progenitors (MPPs). Genetic and molecular studies of HSC self-renewal have identified candidate regulatory factors, including cell-intrinsic regulators, such as transcription factors and cell surface receptors, and cell-extrinsic regulators, such as the bone marrow niche and cytokines. Under certain conditions, such as inflammatory stress, HSCs differentiate into progenitor cells with less ability to self-renew, and they can be stimulated to divide and/or differentiate into all cell types in the peripheral blood [5,6]. Under inflammatory conditions, such as during bacterial infection or sepsis, an apparent expansion of lineage-negative Sca-1+c-Kit+ bone marrow cells (KSL) has been observed [7–11]. HSCs and progenitor cells are involved in the expansion of KSL; and expansion of the KSL population in the bone marrow has been associated with a loss of dormant LT-HSCs, reduced engraftment, and a bias towards myeloid lineage differentiation within that population. The process of the transition of HSCs from dormancy to activity is mediated by type I interferon (IFN) and type II IFN.

Interferon is produced by cells of the immune system in response to challenge by agents, such as viruses, bacteria, and tumor cells. Type I IFNs are induced by the genomes of many RNA viruses during viral infection, and they suppress viral replication and have immunomodulatory activity. IFNs are used clinically to treat viral diseases and malignancies, such as chronic myeloid leukemia (CML)[12]. In addition, under steady state conditions in the absence of infection, a small amount of intrinsic IFN is produced constitutively [13]. Recently, Essers et al. demonstrated that chronic activation of the IFN-α pathway impairs the function of HSCs and

INTECH
open science | open minds

that acute IFN-α treatment promotes the proliferation of dHSCs in vivo. Studies exploring the application of type I IFN to target cancer stem cells are expected [14].

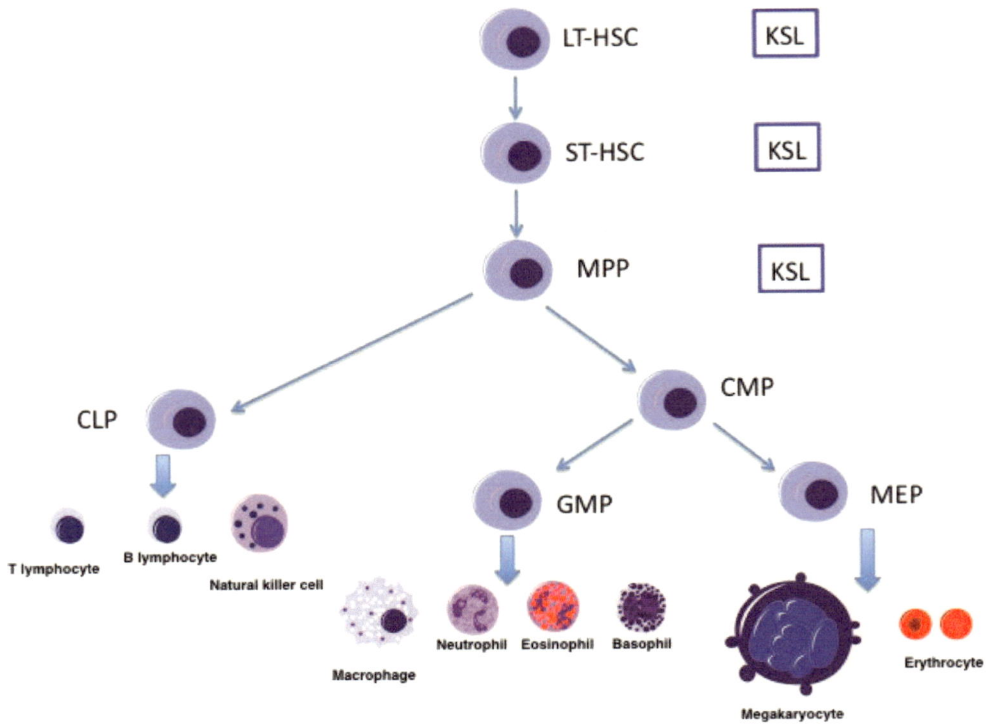

Figure 1. The fates of hematopoietic stem cells. HSCs are capable of self-renewal and differentiation into various lineages of the hematopoietic system to form all types of blood cells. LT-HSCs, ST-HSCs, and MPP are included in the KSL population.

Interferon regulatory factor-2 (IRF-2) is a typical inhibitor of the IFN response. Previously, I showed that IRF-2-/- mouse bone marrow cells contain an enhanced population of Sca-1-positive cells and reduced HSC activity in the KSL fraction [15]. Sato et al. reported that a marked reduction of hematopoietic stem cells in IRF-2-/- mice was dependent on a type I IFN-dependent mechanism [16]. In IRF-2-/- mouse bone marrow, type I IFN signalling is constitutively activated, and interferon-stimulated genes (ISGs) are continuously induced to activate cell cycle progression in HSCs. IRF-2 plays an important role enabling HSCs to maintain their inhibitory status against the type I IFN response.

Bone marrow HSCs are somatic stem cells that have a number of cell surface markers, which allow their prospective identification and isolation with FACS. Mouse models are usually used to examine HSC function in vivo. However, the HSC cell surface markers differ between mice and humans. Mouse HSCs in healthy adult bone marrow show a Lin-Sca-1+c-Kit+CD34-

http://dx.doi.org/10.5772/54689

CD150+CD48- phenotype. In contrast, human HSCs are less well defined but present in the Lin-CD34+CD38- population within cord blood.

Infection or inflammation usually induces IFN production for host defense. Thus, the IFN response is a key factor in switching HSC activity to renewal in the hematopoietic system during the response to environmental conditions. This review will focus on the regulation of HSCs mediated by types I and II IFN.

2. Hematopoietic stem cells in bone marrow

Under homeostatic conditions, HSCs are thought to be quiescent, and they are referred to as long-term reconstituting HSCs (LT-HSC) or dormant HSCs (dHSCs) [3,4]. During homeostasis, dormant HSCs(dHSCs) are consistently inactive with residing bone marrow niches. Homeostatic HSCs (hHSCs) divide and self-renew to generate the progenitors [14]. dHSCs are almost in the G0 phase of cell cycle and found in niches within the cavities of trabecular bone. dHSCs are resistant to anti-proliferative chemotherapy such as 5-fluorouracil (5-FU). Homeostatic HSCs(hHSCs) divide and self-renew. dHSCs and hHSCs exit their niches and undergo self-renewing divisions, followed by maintenance to homeostasis of hematopoietic cell populations. IFN-α induces cell cycle entry of dHSCs and hHSCs, resulting in rapidly dividing active HSCs (aHSC), which can produce progenitors and then differentiate ito mature cell types [14].

3. Type I IFN (IFN-α)

IFNs play a critical role in the regulation of resistance to viral infection and activation of the innate and acquired immune system through IFN-stimulated genes (ISGs). IFNs also have an anti-malignant effect and are important intrinsic cytokines for host defense. To induce ISGs, type I IFN signaling is mediated by activation of the Jak-STAT signaling pathway through the type I IFN receptor (Figure 2). Many RNA viruses and double-stranded RNAs that mimic polyinosinic-polycytidylic acid (poly[I:C]) induce the expression of type I interferons (IFN-α, IFN-β) that bind to the IFNα/β receptor (IFNAR), which mediates Jak-STAT signaling followed by ISG transcription. Treatment with IFN-α induces transcription and cell surface presentation of stem cell marker Sca-1, which is downstream of IFN-α signaling, on hematopoietic stem cells and progenitor cells [14].

The KSL population from IRF-2-/- mouse bone marrow is increased due to the enhancement of Sca-1. However, the CD150+CD48-KSL population is dramatically reduced because of a reduction of the CD150+ population in IRF-2-/- mouse bone marrow cells. IRF-2-/- HSCs are incapable of reconstituting lethally irradiated mice because of constitutive IFN signaling. Chronic exposure to IFN-α has a negative effect on HSC self-renewal. IFN-α induces the activation of the dormant HSC pool. Sato et al. indicated that chronic exposure of HSCs from IRF-2-/- mice to IFN-α resulted in a marked reduction of LT-HSCs and dHSCs[16]. This reduction of LT-HSC activity was rescued in IRF-2-/-IFNAR-/- dKO mice, but not completely, suggesting an IFN-independent mechanism exists in IRF-2-/- mice. Gene expression and transplantation analyses using HSCs from IRF-2-/-IFNAR-/- mice have demonstrated that IRF-2 regulates HSC populations independently of type I IFN signaling.

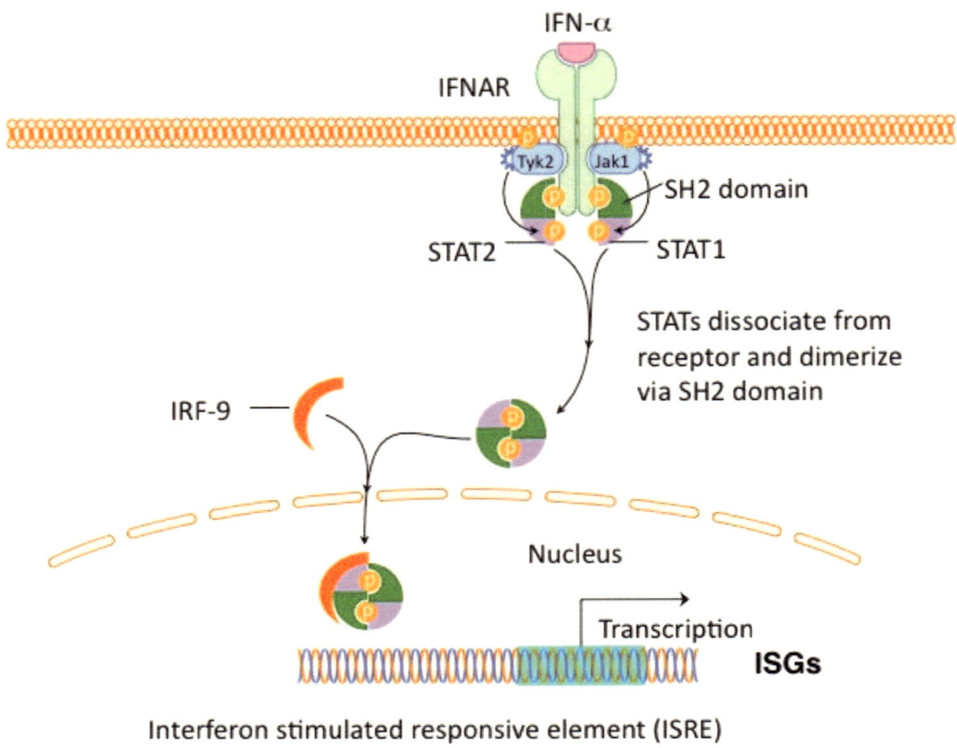

Figure 2. Jak-Stat signaling pathway activated by Interferon. Type I IFN binds to the IFN receptor to stimulate the Jak-STAT signaling pathway. Phosphorylated JAK1and Tyk2 phosphorylate STAT1 and STAT2, which heterodimerize via SH2 domain, and then translocated to nucleolus with IRF-9. The interferon-stimulated gene factor-3 (ISGF3) complex is transferred to the nucleus to bind to the ISRE consensus sequence that induces transcription, resulting in production of ISGs. STAT: Signal Transducers and Activator of Transcription, SH2:Src homology domain

A large quantity of whole bone marrow cells from IRF-2-/- mice can rescue recipients from lethal irradiation [15]. Whole bone marrow cells from IRF-2-/- mice were transplanted into lethally irradiated wild type mice, and KSL or CD150+CD48-Lin- cells from the recipient mice were isolated. Higher doses of bone marrow cells from IRF-2-/- mice could rescue recipients from lethal irradiation, even if the engraftment efficiency was poorer than that of wild-type mice. In addition, in the rescued recipients, the percentage of donor-derived KSL and CD150+CD48-Lin- cells from IRF-2-/- mice was higher compared to that of wild-type mice (Masumi et al., unpublished data).

The number of CD150+CD48-Lin- cells from IRF-2-/- mouse bone marrow present in a recipient wild-type environment is almost comparable to that of wild-type (Masumi et

http://dx.doi.org/10.5772/54689

al., unpublished data). A marked reduction of CD150+CD48-Lin- cells in the bone marrow of IRF-2-/- mice may be due to the impairment of the stem cell niche in IRF-2-/- mouse bone marrow. Essers et al. reported that wild-type HSC in chimeric mice with wild-type hematopoietic cells in an IFNAR-/- stromal environment are still efficiently activated by IFN-α and that IFN-α signaling is not required in the stem cell niche. The CD150+ population in KSL cells may be partly regulated by an IRF-2-dependent stem cell niche environment. In contrast, KSL populations from IRF-2-/- mouse bone marrow in the stromal environment of the recipient wild-type mice still show the phenotype of IRF-2-/- mice. CD150+CD48-KSL HSCs in untreated mice are predominantly in a quiescent intracellular Ki67-negative G0 phase. After injection of poly (I:C), most of these cells exit G0 and enter G1, activating the cell cycle. CD150+CD48-KSL in IRF-2-/- mice continuously enter an active cell-cycle state, which is dependent on chronic type I IFN signaling (Figure 3).

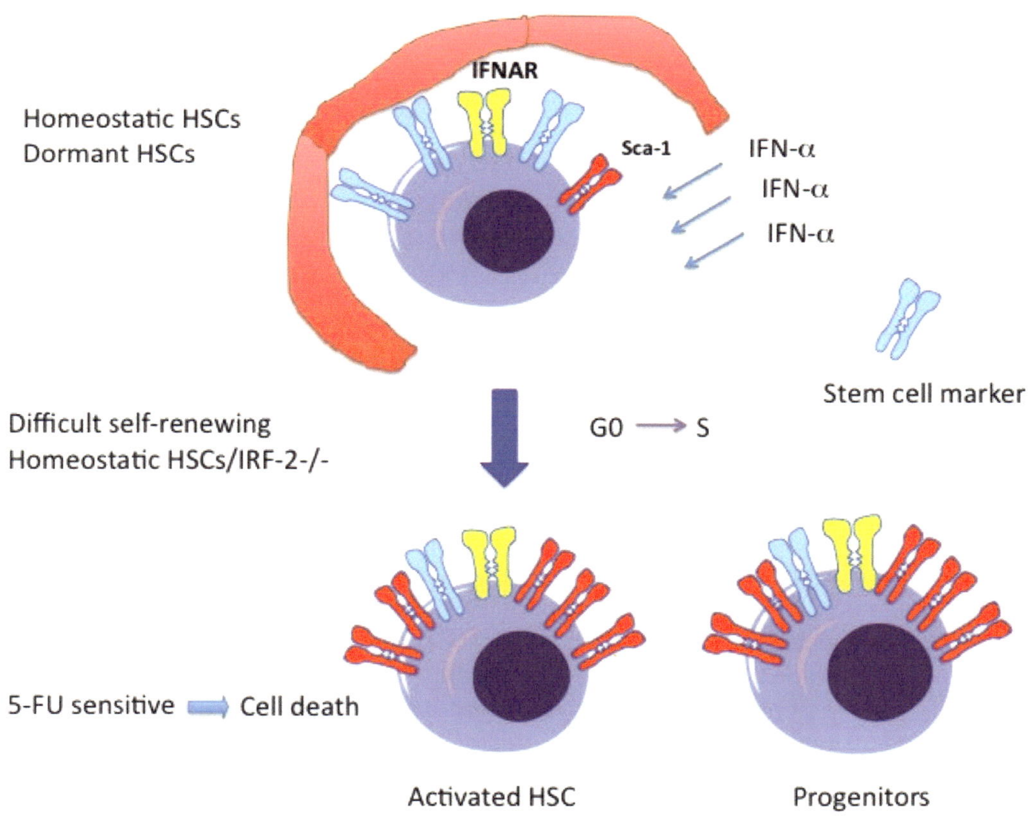

Figure 3. Chronic IFN stress in IRF-2-/- mice. In IRF-2-/- mice bone marrow cells, which cannot abrogate chronic IFN stimulation, Sca-1 is upregulated at the mRNA and protein levels, resulting in enhanced Sca-1 on the cell surface. However, CD150+CD48- cells are almost absent in the KSL of IRF-2-/- mouse bone marrow. Chronic IFN-α stimulation induces cell cycle entry of dormant and homeostatic HSCs as well as progenitors (MPPs).

Most interestingly, several hematologic malignancies, including acute myeloid leukemia (AML) and chronic myeloid leukemia (CML), often contain a small population of malignant cells that show similarities in phenotype and function to normal HSCs. Dormant HSCs, which show the highest self-renewal potential of all HSCs and are almost permanently in the G0 phase of the cell cycle, are resistant to antiproliferative chemotherapy with agents, such as 5-FU. Cancer stem cells (CSCs) are resistant to radiation and antiproliferative chemotherapeutic agents because of their distinctive properties, which seem to be related to their stem cell-like character [17].

AML stem cells are reported to be located at the endosteal region of the bone marrow and are mainly noncycling. The tyrosine kinase inhibitor imatinib mesylate (Gleevec) blocks the constitutively active BCR-ABL kinase produced by the Philadelphia chromosome and is the first example of targeted chemotherapy as it acts against the causative mutation (BCR-ABL) that initiates the disease [18,19]. IFN-α was formerly a first-line treatment for CML, with variable outcomes. Imatinib mesylate has replaced IFN-α owing to its better response rate and fewer side effects. However, intrinsic IFN-α may help the molecular agents that target cancer stem cells.

4. Effect of IFN-γ signaling in bone marrow

IFN-γ plays a critical role in innate immunity to intracellular pathogens, including the ehrlichiae, and is essential for host defense because of its ability to activate macrophages and promote Th1 responses. Many in vivo infectious studies have demonstrated that numerous pathogens induce production of IFN-γ, which mediates host defense. IFN-γ is a critical defense against mycobacteria, as mice with a disrupted IFN-γ receptor gene, IFNGR-1, did not respond to *M. avium* infection [20]; and human patients with mutations in the IFN-γ receptor have died from disseminated *M. avium* infection [21]. However, a lesser known function of IFN-γ is its ability to act as a moderator of hematopoiesis. Several studies have demonstrated that IFN-γ inhibits formation of human bone marrow colonies and the growth of CD34+ bone marrow cells through differentiation and apoptosis [22–26]. In vitro, the addition of exogenous IFN-γ significantly inhibits myeloid and erythroid colony formation in both murine and human systems [22,27–30]. Zhao et al. demonstrated that IFN- is essential for the generation of a unique progenitor population that gives rise to mostly myeloid cells [31]. A colony-forming unit (CFU) assay indicated that treatment with IFN-γ led to a reduction in the number of progenitor cells (Lin-c-Kit+Sca-1-) in bone marrow and enhanced them in the spleen [30]. Sorted KSL cells from IFN-γ-treated mice had greater potential for proliferation and were able to yield higher CFUs compared to control progenitor and KSL cells.

Genes induced by IFN-γ are critical for HSC survival and maintenance via the expansion and proliferation of KSL [20,32]. This observation shows that IFN-γ modulates the differentiation of hematopoietic progenitor cells, resulting in the enhanced production of mature blood cells. Investigations of mycobacterial infection have defined an important role for IFN-γ in myelopoiesis [20,33–35]. Several bacterial and viral infection models have confirmed the effect of

http://dx.doi.org/10.5772/54689

IFN-γ on hematopoietic stem and progenitor cells (HSPCs). A murine in vivo infection model and in vitro human cell culture study showed that *Escherichia coli* stimulates the expansion and mobilization of hematopoietic progenitors (CD34-positive cells in humans and KSL cells in a murine system) [36–39]. Vaccinia virus infection increases immature myelocytes in peripheral blood [10]. These infections trigger the mobilization and differentiation of common myeloid progenitors. Intracellular bacterial infections cause the differentiation of hematopoietic progenitor cells through direct IFN-γ signaling (Figure 4).

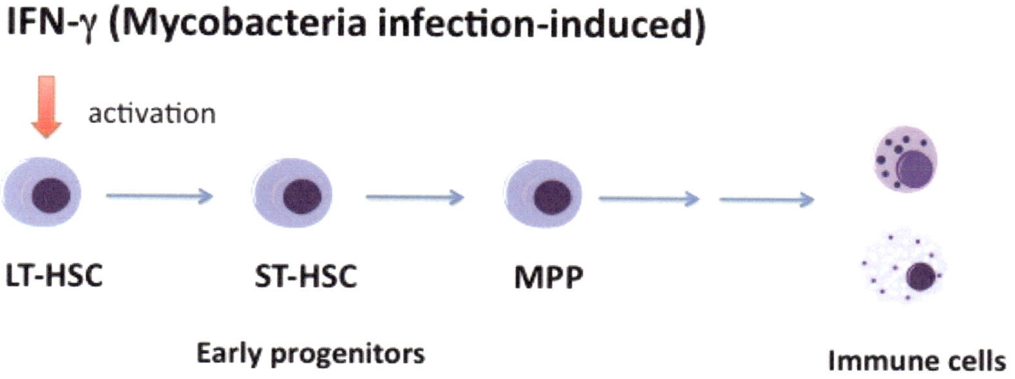

Figure 4. During infection, IFN-γ is produced by several immune cells such as macrophages, NK cells, and lymphocytes. IFN-γ activates HSCs and induces the expansion of MPPs.

Acute bacterial infection (*Ehrlichia muris*) induces myelopoiesis that is dependent on IFN-γ signaling. It has been demonstrated that infection-induced IFN-γ acts on normally quiescent HSCs, leading to transient activation and promoting an expedited innate immune response [7,8]. Infection of mice with the malaria parasite *Plasmodium chabaudi* caused anemia and a decrease in erythrocytes. Malarial infection induces IL-7Ra+c-Kit^{hi} progenitors, which generate myeloid cells to decrease parasitemia in vivo. Lin-IL-7Ra+c-Kit^{hi} cells contribute to the clearance of malaria-infected erythrocytes in vivo. The infection of ifngr-/- mice with *P. chabaudi* did not increase the number of Lin-IL-7Ra+c-Kit^{hi} atypical progenitors [34]. Thus, IFN-γ signaling is important for hematopoiesis and the innate immune system's response to acute infection.

Chronic mycobacterial infection also induces the proliferation of HSCs. *M. avium* infection seems to induce an increase in the proliferative fraction of primitive LT-HSCs and a substantial increase in the number of early-committed progenitors. IFN-γ is strongly upregulated during *M. avium* infection. HSCs from Stat-1-/- and ifngr-/- mice infected with *M. avium* showed an impaired proliferative response by the LT-HSC population, indicating that the HSC response to *M. avium* infection is dependent on IFN-γ signaling [33]. The effects of both in vitro and in vivo IFN-γ treatment on KSL expansion were greatly reduced in Stat-1-/- and ifngr-/- mice compared to those in control mice [31]. Enhanced incorporation of BrdU into LT-HSCs from IFN-γ-treated mice and a modest impairment of the engraftment of whole bone marrow cells in IFN-γ-treated mice were detected [33]. IFN-γ directly affects and modulates the proliferation and mobilization of HSCs.

4.1. Human diseases and IFN-γ signaling

Human bone marrow failure syndromes, such as myelodysplastic syndromes (MDS), aplastic anemia (AA) and paroxysmal nocturnal hemoglobinuria (PNH) are characterized by a defect of stem and progenitor cell populations. The gene expression analysis of human CD34 cells treated with IFN-γ indicated results that were similar to those of CD34 cells that are derived from patients with bone marrow failure syndromes, according to Zheng et al. [40]. IFN-γ negatively modulates the self-renewal of CD34+CD38- stem cells and promotes differentiation of CD34+ cord blood cells expanded in vitro [26]. IFN-γ is an important modulator of myeloid differentiation in myelocytopenias and pathogenesis.

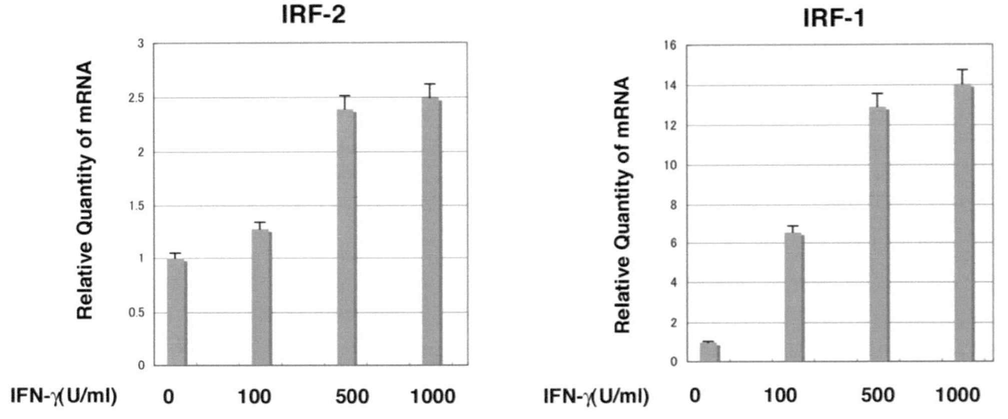

Figure 5. IFN-γ stimulation induces expression of IRF-2 and IRF-1 in mouse hematopoietic stem/progenitor cells. Bone marrow cells from mice injected with 5-FU (150 mg/kg) were cultured in the presence of IFN-γ at the indicated concentrations. Real-time PCR was performed and the relative amounts of mRNA from the IRF-1 and IRF-2 genes were quantified with a β-actin primer [44].

4.2. IFN-γ signaling and megakaryopoiesis

IFN-γ has been reported to induce production of megakaryocytes and platelets [41,42]. A previous study reported that IFN-γ accelerates megakaryocyte differentiation and IRF-1 and IRF-2 are IFN-γ-inducible factors that enhance megakaryopoiesis [43–45]. Huang et al. reported that GATA-1 promotes megakaryopoiesis through the activation of IFN-γ/STAT-1 signaling, of which IRF-1 is a downstream target. IFN-γ enhanced colony formation and proliferation of megakaryocytes, but not their maturation [43]. When IFN-γ was stimulated with bone marrow cells isolated from 5-FU-injected mice, IRF-2 expression was increased by around 80-100%, but its induction level was much lower than that of IRF-1 (Fig. 5). Transduction of IRF-2 into mouse bone marrow stem/progenitor cells induces megakaryocyte differentiation, as we demonstrated with a colony formation assay [44]. Upregulation of IRF-2 by IFN-γ treatment leads to CD41 expression, resulted in megakaryopoiesis [44]. IRF-2 plays an important role for hematopoiesis mediated by each type I and type II IFN response. However, the number of peripheral blood platelets is comparable between control mice and mice transplanted with IRF-2-transduced bone marrow stem cells/progenitors. IRF-2 enhances megakaryocyte colony formation, not its maturation, suggesting its cooperation with other factors for maturation [44, 45].

5. Conclusion

HSCs might be driven to exhaustion by IFN stimulation. IFN-γ modulates HSCs differentiation into progenitors for host defense against extrinsic pathogen, inflammation, and other immune responses. IFN-α/β stimulates quiescent/dormant HSCs to proliferate, leading to HSC exhaustion. Under conditions of chronic IFN stimulation including inflammation, infection, and IRF-2 deficiency, not only hematopoietic stem cells, but also cancer stem cells change from their dormant status and enter the cell cycle to proliferate. Many anticancer therapies, such as 5-FU, can disrupt proliferating cells, but not quiescent cells. Furthermore, cancer cells sometimes become resistant to drugs with molecular targets, such as imatinib. Targeting specific intrinsic IFN signaling in conjunction with the use of anticancer drugs may be a useful approach for abrogating cancer stem cells.

Acknowledgements

I thank Drs. I. Hamaguchi, K. Takizawa, T. Mizukami, M. Kuramitsu, H. Momose, and K. Yamaguchi for their experimental support and useful discussions. I wish to also thank Ms. K. Furuhata for cell sorting and Ms. M. Tsuruhara and Ms. K.Araki for molecular technical assistance. This work was supported in part the Japan Society for the Promotion of Science and the Ministry of Education, Science, Sports, and Culture of Japan.

Author details

Atsuko Masumi*

Address all correspondence to: amasumi@nih.go.jp

Department of Safety Research on Blood and Biological Products, National Institute of Infectious Diseases, Tokyo, Japan

References

[1] Ema, H, Morita, Y, Yamazaki, S, Matsubara, A, Seita, J, Tadokoro, Y, Kondo, H, Takano, H, & Nakauchi, H. Adult Mouse Hematopoietic Stem Cells: Purification and Single-Cell Assays. Nature Protocols (2006). , 1, 2979-2987.

[2] Orkin, S. H, & Zon, L. I. Hematopoiesis: An Evolving Paradigm for Stem Cell Biology. Cell (2008). , 132, 631-644.

[3] Arai, F, & Suda, T. (2007). Maintenance of Quiescent Hematopoietic Stem Cells in the Osteoblastic Niche. Annals of the New York Academy of Sciences 2007;, 1106, 41-53.

[4] Orford, K. W, & Scadden, D. T. Deconstructing Stem Cell Self-Renewal: Genetic Insights into Cell-Cycle Regulation. Nature Reviews Genetics (2008). , 9, 115-128.

[5] Majeti, R, Park, C. Y, & Weissman, I. L. Identification of a Hierarchy of Multipotent Hematopoietic Progenitors in Human Cord Blood. Cell Stem Cell (2007). , 1, 635-645.

[6] Mcdermott, S. P, Eppert, K, Lechman, E. R, Doedens, M, & Dick, J. E. Comparison of Human Cord Blood Engraftment between Immunocompromised Mouse Strains. Blood (2010). , 116, 193-200.

[7] MacNamara KCJones M, Martin O, Winslow GM. Transient Activation of Hematopoietic Stem and Progenitor Cells by IFN-gamma during Acute Bacterial Infection. PLoS One (2011). e28669. DOI:10.1371/journal.pone.0028669.

[8] MacNamara KCOduro K, Martin O, Jones DD, McLaughlin M, Choi K, D.L. Borjesson, G.M. Winslow. Infection-induced Myelopoiesis during Intracellular Bacterial Infection is Critically Dependent Upon IFN-Gamma Signaling. Journal of Immunology 186; (2011). , 2011, 1032-1043.

[9] Scumpia, P, Kely-scumpia, K, Delano, M, Weinstein, J, Cuenca, A, Al-quran, S, Bovio, I, Akira, S, Kumagai, Y, & Moldawer, L. Bacterial Infection Induces Hematopoietic Stem and Progenitor Cell Expansion in the Absence of TLR Signaling. Journal of Immunology (2010). , 184, 2247-2251.

[10] Singh, P, Yao, Y, Weliver, A, Broxmeyer, H, Hong, S, & Chang, C. Vaccinia Virus Infection Modulates the Hematopoietic Cell Compartments in the Bone Marrow. Stem Cells (2008). , 26, 1009-1016.

[11] Welner, R. S, Pelayo, R, Nagai, Y, Garrett, K. P, Wuest, T. R, Carr, D. J, Borghesi, L. A, Farrar, M. A, & Kincade, P. W. Lymphoid Precursors are directed to Produce Dendritic Cells as a Result of TLR9 Ligation during Herpes Infection. Blood (2008). , 112, 3753-3761.

[12] Kiladjian, J. J, Mesa, R. A, & Hoffman, R. The renaissance of interferon therapy for the treatment of myeloid malignancies, Blood (2011). , 117(2011), 4706-4715.

[13] Taniguchi, T, Ogasawara, K, Takaoka, A, & Tanaka, N. IRF Family of Transcription Factors as Regulators of Host Defense. Annual Review Immunology (2001). , 19, 623-655.

[14] Essers, M, Offner, S, Blanco-bose, W, Waibler, Z, Kalinke, U, Duchosal, M, & Trumpp, A. IFNa Activates Dormant Hematopoietic Stem Cells In Vivo. Nature (2009). , 458, 904-909.

[15] Masumi, A, Miyatake, S, Kohno, T, & Matsuyama, T. Interferon Regulatory Factor-2 Regulates Hematopoietic Stem Cells in Mouse Bone Marrow. InTech-Open (2012). , 2012, 91-112.

[16] Sato, T, Onai, N, Yoshihara, H, Arai, F, Suda, T, & Ohteki, T. Interferon Regulatory Factor-2 Protects Quiescent Hematopoietic Stem Cells from Type 1 Interferon-Dependent Exhaustion. Nature Medicine (2009). , 15, 696-701.

[17] Van Der Wath, R. C, Wilson, A, Laurenti, E, Trumpp, A, & Lio, P. Estimating Dormant and Active Hematopoietic Stem Cell Kinetics through Extensive Modeling of Bromodeoxyuridine Label-retaining Cell Dynamics. PLoS One (2009). e6972. DOI:10.1371/journal.pone.0006972.

[18] Goldman, J. M. Advances in CML. Clinical Advances Hematology & Oncology (2007). , 5, 270-272.

[19] Goldman, J. M. Chronic Myeloid Leukemia Stem Cells: Now on the Run. Journal of Clinical Oncology (2009). , 27, 313-314.

[20] Feng, C. G, Weksberg, D. C, Taylor, G. A, Sher, A, Goodell, M. A, & The, p. GTPase Lrg-47 (Irgm1) Links Host Defense and Hematopoietic Stem Cell Proliferation. Cell Stem Cell (2008). , 2, 83-89.

[21] Horwitz, M, Uzel, G, Linton, G, Miler, J, Brown, M, Malech, H, & Holland, S. Persistent Mycobacterium Avium Infection Following Nonmyeloablative Allogenic Peripheral Blood Stem Cell Transplantation for Interferon-g Receptor-1 Deficiency. Blood (2003). , 102, 2692-2694.

[22] Hwang, J. H, Kim, S. W, Lee, H. J, Yun, H. J, Kim, S, & Jo, D. Y. Interferon Gamma has Dual Potential in Inhibiting or Promoting Survival and Growth of Hematopoietic Progenitors: Interactions with Stromal Cell-derived Factor 1. International Journal of Hematolology (2006). , 84, 143-150.

[23] Raefsky, E. L, Platanias, L. C, Zoumbos, N. C, & Young, N. S. Studies of Interferon as a Regulator of Hematopoietic Cell Proliferation. Journal of Immunology (1985). , 135, 2507-2512.

[24] Selleri, C, Maciejewski, J. P, Sato, T, & Young, N. S. Interferon-gamma Constitutively Expressed in the Stromal Microenvironment of Human Marrow Cultures Mediates Potent Hematopoietic Inhibition. Blood (1996). , 87, 4149-4157.

[25] Snoeck, H. W, Van Bockstaele, D. R, Nys, G, Lenjou, M, Lardon, F, Haenen, L, Rodrigus, I, Peetermans, M. E, & Berneman, Z. N. Interferon-gamma Selectively Inhibits Very

Primitive CD342+CD38- and Not More Mature CD34+CD38+ Human Hematopoietic Progenitor Cells. Journal of Experimental Medicine (1994). , 180, 1177-1182.

[26] Yang, L, Dybedal, I, Bryder, D, Nilsson, L, Sitnicka, E, Sasaki, Y, & Jacobsen, S. E. IFN-gamma Negatively Modulates Self-renewal of Repopulating Human Hemopoietic Stem Cells. Journal of Immunology (2005). , 174, 752-757.

[27] Maciejewski, J, Selleri, C, Anderson, S, & Young, N. S. Fas Antigen Expression on CD34+ Human Marrow cells is Induced by Interferon Gamma and Tumor Necrosis Factor Alpha and Potentiates Cytokine-Mediated Hematopoietic Suppression In Vitro. Blood (1995). , 85, 3183-3190.

[28] Sato, T, Selleri, C, Young, N, & Maciejewski, J. Hematopoietic Inhibition by Interferon-gamma is Partially Mediated Through Interferon Regulatory Factor-1. Blood (1995). , 86, 3373-3380.

[29] Zoumbos, N. C, Djeu, J. Y, & Young, N. S. Interferon is the Suppressor of Hematopoiesis Generated by Stimulated Lymphocytes In Vitro. Journal of Immunology (1984). , 133, 769-774.

[30] Zoumbos, N. C, Gascon, P, Djeu, J. Y, & Young, N. S. Interferon is a Mediator of Hematopoietic Suppression in Aplastic Anemia In Vitro and Possibly In Vivo. Proceedings of the National Academy of Sciences of the United States of America (1985). , 82, 188-192.

[31] Zhao, X, Ren, G, Liang, L, Ai, P. Z, Zheng, B, Tischfield, J. A, Shi, Y, & Shao, C. Brief Report: Interferon-gamma Induces Expansion of Lin(-)Sca-1(+)C-Kit(+) Cells. Stem Cells (2010). , 28, 122-126.

[32] Hartner, J. C, Walkley, C. R, Lu, J, & Orkin, S. H. ADAR1 is Essential for the Maintenance of Hematopoiesis and Suppression of Interferon Signaling. Nature Immunology (2009). , 10, 109-115.

[33] Baldridge, M. T, King, K. Y, Boles, N. C, Weksberg, D. C, & Goodell, M. A. Quiescent Haematopoietic Stem cells are Activated by IFN-gamma in Response to Chronic Infection. Nature (2010). , 465, 793-797.

[34] Belyaev, N. N, Brown, D. E, Diaz, A. I, Rae, A, Jarra, W, Thompson, J, Langhorne, J, & Potocnik, A. J. Induction of an IL7-R(+)c-Kit(hi) Myelolymphoid Progenitor Critically Dependent on IFN-gamma Signaling during Acute Malaria. Nature Immunology (2010). , 11, 477-485.

[35] Murray, P. J, Young, R. A, & Daley, G. Q. Hematopoietic Remodeling in Interferon-gamma-deficient Mice Infected with Mycobacteria. Blood (1998). , 91, 2914-2924.

[36] Kim, J. M, Oh, Y. K, Kim, Y. J, Youn, J, & Ahn, M. J. Escherichia Coli Up-regulates Proinflammatory Cytokine Expression in Granulocyte/Macrophage Lineages of CD34 Stem Cells Via Homodimeric NF-kappaB. Clinical and Experimental Immunology (2004). , 50.

[37] Quinton, L. J, Nelson, S, Boe, D. M, Zhang, P, Zhong, Q, Kolls, J. K, & Bagby, G. J. The Granulocyte Colony-Stimulating Factor Response After Intrapulmonary and Systemic Bacterial Challenges. Journal of Infectious Diseases (2002). , 185, 1476-1482.

[38] Shahbazian, L. M, Quinton, L. J, Bagby, G. J, Nelson, S, Wang, G, & Zhang, P. Escherichia Coli Pneumonia Enhances Granulopoiesis and the Mobilization of Myeloid Progenitor Cells into the Systemic Circulation. Critical Care Medicine (2004). , 32, 1740-1746.

[39] Zhang, P, Nelson, S, Bagby, G. J, & Siggins, R. nd, Shellito JE, Welsh DA. The Lineage-c-Kit+Sca-1+ Cell Response to Escherichia Coli Bacteremia in Balb/c mice. Stem Cells (2008). , 26, 1778-1786.

[40] Zheng, C, Li, L, Haak, M, Bros, B, Frank, O, Giehl, M, Fabarius, A, Schatz, M, Weisser, A, Lorentz, C, Gretz, N, Hehlmann, R, Hochhaus, A, & Seifarth, W. Gene Expression Profiling of CD34+ Cells Identifies a Molecular Signature of Chronic Myeloid Leukemia Blast Crisis. Leukemia (2006). , 20, 1028-1034.

[41] Muraoka, K, Tsuji, K, Yoshida, M, Ebihara, Y, Yamada, K, Sui, X, Tanaka, R, & Nakahata, T. Thrombopoietin-independent Effect of Interferon-gamma on the Proliferation of Human Megakaryocyte Progenitors. British Journal of Haematology (1997). , 98, 265-273.

[42] Tsuji-takayama, K, Tahata, H, Harashima, A, Nishida, Y, Izumi, N, Fukuda, S, Ohta, T, & Kurimoto, M. Interferon-gamma Enhances Megakaryocyte Colony-stimulating Activity in Murine Bone Marrow Cells. Journal of Interferon & Cytokine Research (1996). , 16, 701-708.

[43] Huang, Z, Richmond, T, Munteen, A, Barber, D, Weiss, M, & Crispino, J. Stat1 Promotes Megakaryopoiesis Downstream of GATA-1 in Mice. Journal of Clinical Investigation (2007). , 117, 3890-3899.

[44] Masumi, A, Hamaguchi, I, Kuramitsu, M, Mizukami, T, Takizawa, K, Momose, H, Naito, S, & Yamaguchi, K. Interferon Regulatory Factor-2 Induces Megakaryopoiesis in Mouse Bone Marrow Hematopoietic Cells. FEBS Letters (2009). , 583, 3493-3500.

[45] Masumi, A. The Role for Interferon Regulatory Factor-2 on Mouse Hematopoietic Stem Cells in an Inflammation State. Inflammation and Regeneration (2010). , 30, 531-535.

Chapter 6

Stem Cell Predictive Hemotoxicology

Holli Harper and Ivan N. Rich

Additional information is available at the end of the chapter

http://dx.doi.org/10.5772/54430

1. Introduction

Stem cells, those elusive entities that have the capacity for producing, maintaining and reconstituting the integrity of a biological system, also demonstrate the potential to predict partial or life-threatening damage in response to drugs, environmental compounds and other agents. It is ironic however, that in the animal or human, prior to the manifestation of such potential biological damage most, if not all of the stem cells might have been eradicated. To predict possible damage, surrogate *in vitro* stem cell assays have been developed that utilize specific properties and characteristics that divulge, through a measured response, how the system will react to different agents.

In vitro assays that detect toxicity to stem cells of a biological system allow potentially life-threatening damage to be predicted prior to human clinical trials taking place and environmental agents from causing harm. Discussions about stem cells usually focuses not on primary cells, but rather on embryonic stem (ES) cells and induced pluripotent stem (iPS) cells, their ability to produce virtually any type of functional cell and their use in cellular therapy and regenerative medicine. The ES and iPS types of stem cells are, in fact, the least understood of all stem cell types. For many companies investigating or considering using these stem cells routinely as surrogate *in vitro* models for toxicity testing many questions remain, including (1) what is the relevance of these cells, (2) how do they compare with primary stem cell populations, and (3) can they be validated? In many cases, it is not the stem cells themselves, but rather the cells derived from ES and iPS cells that are of interest. Several companies already produce ES- or iPS-derived cardiomyocytes, hepatocytes, neural cells and many other cell types not only for toxicity testing, but also for basic research, cellular therapy and regenerative medicine.

How can stem cells be used to predict toxicity? The answer to this question lies in the characteristics and properties of stem cells and how they respond to different situations.

To understand this better, stem cell systems can be divided into "definitive" and "non-definitive" systems as illustrated in Fig. 1. Definitive stem cell systems are responsible for maintaining a specific biological system. They can be divided into continuously proliferating systems such as the blood-forming or lympho-hematopoietic system, the gastrointestinal system, hair and skin, reproductive organs and cells of the eye cornea. Although not necessarily a continuously proliferating system, the mesenchymal stem cell (MSC), also called the multipotent mesenchymal stromal cell [1] system can been included, because in culture, the MSCs proliferate and can be passaged over a long period of time. Definitive stem cells systems can also demonstrate partial proliferation. These include, but are not limited to, the liver, lung, kidney, heart, pancreas, and the neural/neuronal system. From a toxicological viewpoint, these are not usually considered stem cell systems. Yet, the different types of lineage cells present in these organs and the ability to maintain a specific cell mass has all the intricacies of a stem cell system, especially during development, even though the cell turnover in the adult may be very low. Non-definitive stem cells systems are represented by the ES and iPS cell systems, which can, theoretically, give rise to any of the definitive stem cell systems. Indeed, it is a prerequisite that the production of functionally, mature cells from ES or iPS cells first pass through a definitive stem cell compartment.

2. Stem cell characteristics and properties used for toxicity testing

Stem cells of primary, definitive systems always represent a very small proportion of the tissue or organ cellularity. This proportion is between 0.1 and 0.01% or less. The basic definition of a stem cell is that it possesses the capacity for self-renewal. In fact, stem cell systems are usually termed self-renewal cell systems, meaning that one stem cell can produce two daughter cells that are exact replicas of the parent. However, self-renewal is a difficult property to measure. The capacity for either serial *in vivo* repopulation or *in vitro* serial re-plating is considered a property of stem cells that implicates not only the presence of stem cells, but also their self-renewal capability. The fact that serial *in vitro* re-plating or *in vivo* repopulation cannot be performed ad infinitum is not only an indication for a stem cell hierarchy [2-7], but also for an alternative hypothesis to stem cell self-renewal. This hypothesis states that tissues and organs are endowed with a specific number of stem cells. Once used up, the system ceases to function [8]. Regardless of the hypothesis, this important property can be utilized in a toxicological setting by employing secondary re-plating technology. This allows not only the presence of residual stem cells to be detected that have not been affected by a compound, but any change in sensitivity to a compound that might be important during repeated dose administration.

Stem cells have two other important properties that can be applied to toxicity testing. The first is that they are undifferentiated. The second is that stem cells proliferate. Stem cells can be "determined" into one or more lineages of mature functional cells. When a stem cell becomes determined, it ceases to be a stem cell and becomes a progenitor cell that proliferates and differentiates. The fact that stem cells can be induced to differentiate means that whatever happens at the stem cell level will ultimately affect all downstream events. These characteristics enable stem cells to be the most important predictors of potential toxicity.

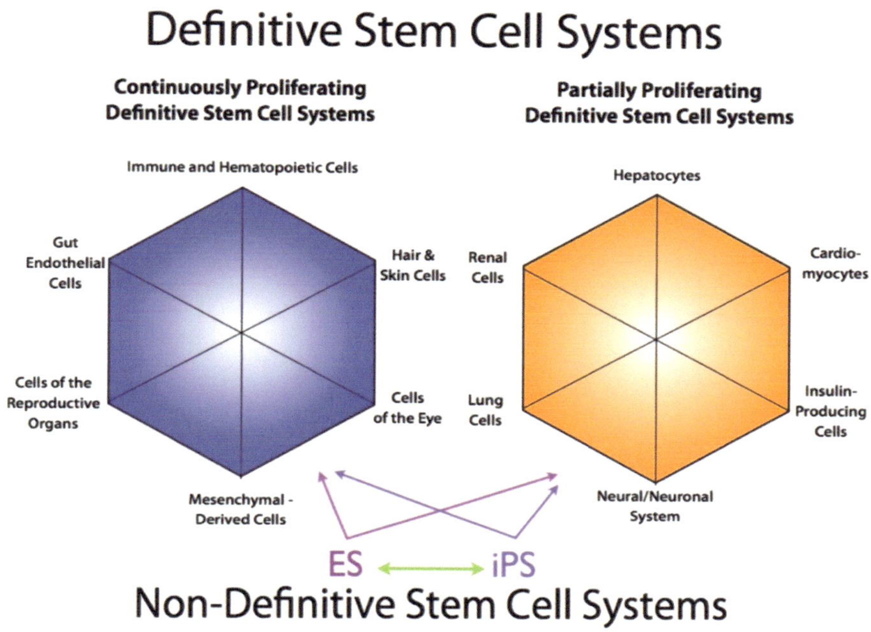

Figure 1. Definitive and Non-Definitive Stem Cell Systems. Definitive stem cell systems can be further divided into continuously and partially proliferating systems. Non-definitive stem cell systems such as embryonic stem cells and induced pluripotent stem cells can produce definitive stem cell systems, which in turn, give rise to mature functional cells.

All definitive stem cell systems have a common organization shown in Fig. 2. There is a continuum of stem cells within the stem cell compartment that exhibit different degrees of primitiveness or "stemness", which in turn, implies changing proliferating potential or potency as a stem cell moves through the compartment to the point of determination. These characteristic properties actually provide the information that allows stem cells to be predictors of potential toxicity. Once a stem cell becomes a progenitor cell, proliferation continues and actually increases for a certain time so that the compartment can be amplified, until it ceases completely and the differentiation and maturation processes takes over. These changes have important implications for the types of assays that can be used *in vitro* to detect potential toxicity.

From Fig. 2, it is clear that proliferation occurs prior to differentiation. Although there is considerable overlap between proliferation and differentiation, they are two separate processes that cannot be measured using the same assay readout. Since stem cells only proliferate, it follows that a proliferation assay is required to detect the presence and response of stem cells to a compound or agent. Using a differentiation assay to detect the effect of a compound or agent that targets one or more steps in the proliferation process can influence the interpretation and conclusion of the results. This can have far-reaching consequences on the decision to move forward with the development of a new drug candidate.

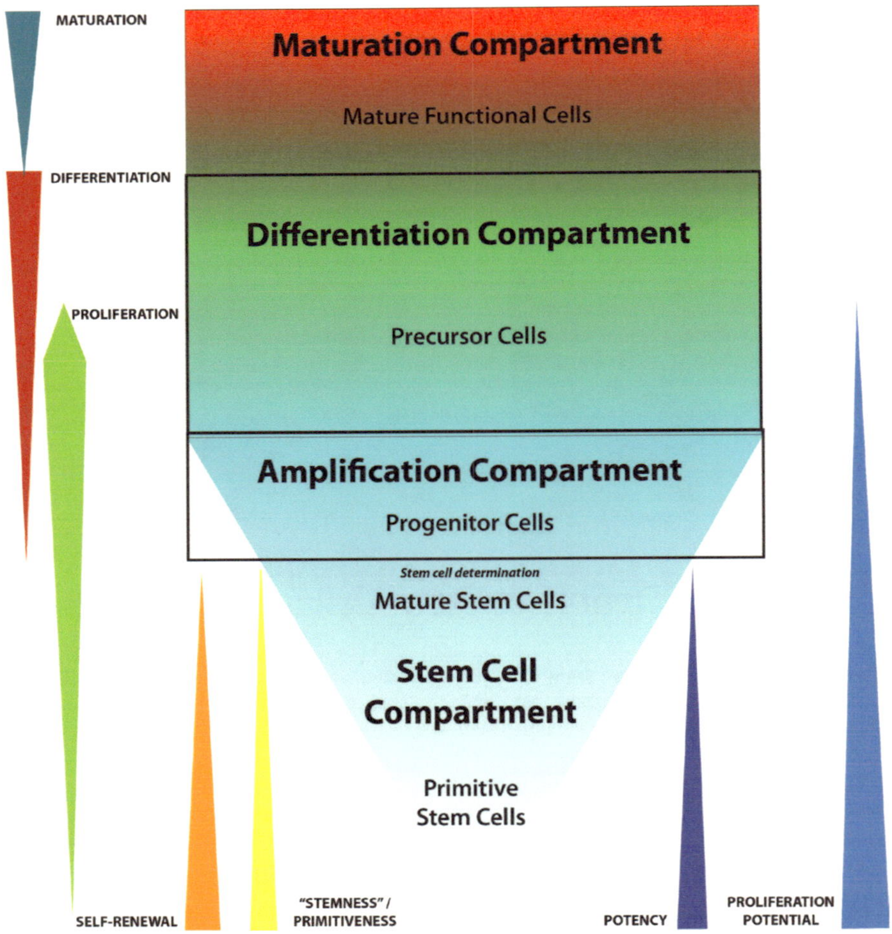

Figure 2. The Common Organization of Definitive Stem Cell Systems

Toxicity represents between 30-40% of the drug attrition rate [9,10]. It is therefore not surprising that biopharmaceutical companies are eager to employ assays that allow early prediction of toxicity prior to starting human clinical trials. Once the drug discovery phase has been concluded, the drug development phase begins (Fig. 3) by screening thousands of compounds in a battery of tests to determine absorption, distribution, metabolism and excretion (ADME) as well as preliminary toxicity (ADME/Tox). Many of the ADME/Tox assays as well as those in the lead optimization phase use transformed cell lines as cell targets, such as the NCI60 cell line panel [11,12]. Once these tests have whittled down the number of possible drug candidates, pre-clinical animal models are used. Neither cell lines (even if they are of human origin) nor animal testing provide good extrapolation to the human situation. It is not uncommon for unexpected results or toxicity to rear its head during animal studies because of the lack of predictive information obtained during previous screening and testing [13]. Many published articles have dealt with this problem, one of the most notable being the monograph on Toxicity Testing in the 21st Century [14]. Despite the goals of the drug development pipeline and the considerable effort being undertaken by regulatory agencies [15-20] to determine the effect of environmental agents on human cells, interpretation and conclusions often fall short due to lack of understanding of the mechanism of action of the molecule, incorrect assay readout and/or incorrect target cell, to name but a few reasons. If the goal is to determine the effect on human cells, then mouse, rat, dog or even non-human primate cells will not provide the required information; human cells must be used. It goes without saying that drug development or testing xenobiotic agents cannot be performed on human subjects. It is for this reason why surrogate *in vitro* assays using primary human cells obtained from donors under the auspices of regulatory controlled internal review boards (IRBs) provide the best alternative. However, even under these circumstances, detailed knowledge of the biological system under study is necessary in order to interpret and make conclusions in the most objective manner.

Of all the biological systems of the body, the one most studied is also one of the systems that is given the least priority with respect to toxicity. The blood-forming or hematopoietic system and the gastrointestinal system are two continuously proliferating systems that are expected to be dramatically affected by anti-proliferating agents such as anti-cancer drugs. As a result, the only relevant questions are (a) how severe would toxicity be, and (b) would use of the drug provide a favorable therapeutic index?

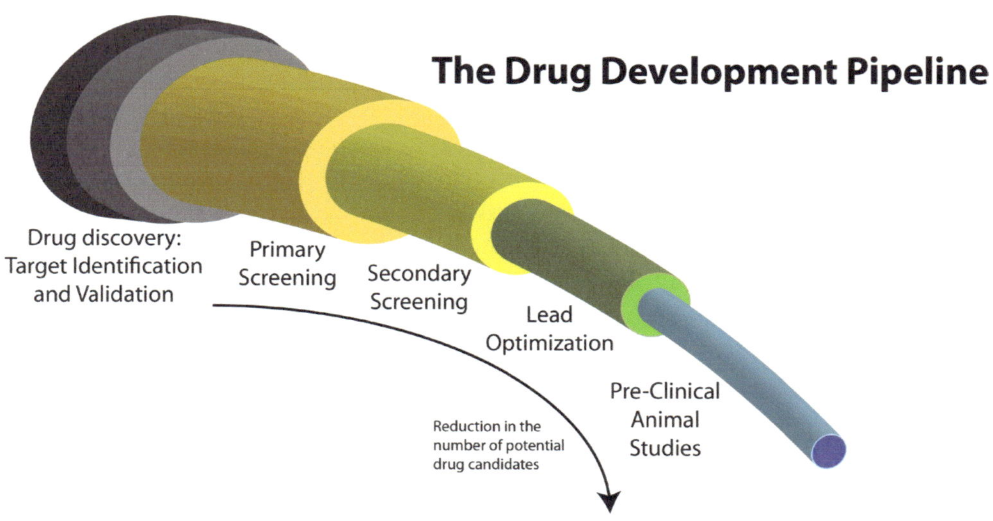

Figure 3. The Major Stages of the Drug Development Pipeline

Hemotoxicity testing is traditionally performed during the last stage of drug development, namely pre-clinical animal testing. Circulating blood parameters are measured and at necropsy, bone marrow, spleen and even liver hematopathology are performed. Primary stem and progenitor cells cannot be morphologically identified. Morphological identification of cells is only possible once the cells start to differentiate and mature. Consequently, traditional hemotoxicity testing provides little, if any, predictive value since most of the toxic effects have occurred on more primitive cells.

Much of our knowledge about the characterization, properties and responses of hematopoietic stem cells and the system as a whole has been provided through the use of drugs and other agents (e.g. radiation) using both *in vivo* and later, *in vitro* assays. The information obtained has allowed the organization and hierarchy within the different compartments of the hematopoietic and lymphopoietic systems to be elucidated. By utilizing the knowledge that has accrued over more than six decades, analysis of the lympho-hematopoietic stem and progenitor cells provide the highest degree of predictive toxicity of any biological system.

3. The Colony-Forming Cell (CFC) assay and ECVAM studies

In 1966, Bradley and Metcalf in Melbourne, Australia [21] and Pluznik and Sachs in Rehovot, Israel, [22] independently published what is now known as the colony-forming unit (CFU) or colony-forming cell (CFC) assay. In its original form, mouse bone marrow target cells were suspended in agar containing a conditioned medium that we now know contained granulocyte-macrophage colony stimulating factor or GM-CSF as well as other soluble factors. In the semi-solid medium, the cells underwent proliferation and later differentiation to produce colonies of cells that were identified either as neutrophils, macrophages or a combination of the two cell types. The number of colonies counted under an inverted microscope was proportional to both the number of cells plated and the dose of the conditioned medium added. In the same year, Cole and Paul [23] in Glasgow, Scotland reported the first *in vitro* suspension culture of murine erythropoietic cells from the yolk sac and fetal liver. Culture of erythropoietic cells under clonal conditions did not occur until 1971, when the Axelrad group [24] in Toronto, Canada, demonstrated that erythroid colonies could be produced using a plasma clot technique. In 1974, Iscove and colleagues [25] in Basel, Switzerland introduced the methylcellulose CFC assay that is still used today. Since that time, colony assays have been developed to detect multiple cell populations of every blood cell lineage, including several different stem cell populations. In addition, conditioned media has been replaced with recombinant growth factors and cytokines.

In Section 2, emphasis was placed on the importance between proliferation and differentiation. The cell populations detected using the CFC assay must all be proliferating populations, otherwise the production of colonies would not occur. However, to identify the type of colony, the *in vitro* culture must be allowed to proceed long enough so that the cells produced can themselves be identified as being derived from a morphologically unidentifiable stem, progenitor or precursor cell, all of which are capable of proliferation, but to different extents.

As mentioned above, although proliferation is required to produce colonies, the number of cells produced as a quantitative measure of proliferation cannot be ascertained. Although proliferation is assumed, the CFC assay actually detects differentiation ability or potential. This has important consequences for toxicity testing.

Over several years, the European Center for the Validation of Alternative Methods (ECVAM) undertook a series of studies in which a number of drugs and chemicals were tested using the CFC assay. These studies are noteworthy because they represented the first attempt to validate a prediction model for assessing the maximum tolerated dose (MTD, equivalent to the IC90 value) for drugs that induce neutropenia [26,27] using the CFC assay. The studies were performed in different laboratories and were later extended to compounds that caused thrombocytopenia [28]. Potential neutropenia was detected by the effect on the granulocyte-macrophage colony-forming cell or GM-CFC (also called CFC or CFC-GM), while thrombocytopenia was detected by the effect on the megakaryocyte colony-forming cell or Mk-CFC (also called CFC-Mk). A decrease or inhibition in the number of colonies counted derived from GM-CFC or Mk-CFC predicted a reduction in neutrophils or platelets in the circulation. The authors demonstrated that the model could correctly predict the MTD of 20 out of 23 drugs tested (87% predictive rate).

There are two points worth emphasizing. First, not all compounds will produce an estimated IC90 value and may not even produce an IC50 value, when tested using the CFC assay. Does that mean that these compounds will not produce neutropenia or thrombocytopenia? It is interesting to note that the same CFC assay that is used to predict toxicity causing neutropenia or thrombocytopenia, is also used in an opposite manner to predict time to neutrophil or platelet engraftment after bone marrow, mobilized peripheral blood or umbilical cord blood stem cell transplantation for cellular therapy [39-31]. In either case, the GM-CFC or MK-CFC populations provide no information on the response of the more sensitive and more important stem cells. After all, it is the hematopoietic stem cells that give rise to both of these populations. This leads to the second point, namely that many compounds target one or more steps in the proliferation process, either at a molecular and/or cellular level. Although both GM-CFC and Mk-CFC populations are proliferating progenitor cell populations, they are not always the primary targets. When a compound affects more than one lineage, the primary effect is not on those lineages individually, but on the common cell that gives rise to those lineages, namely the stem cells [32]. From a practical viewpoint, however, the CFC assay posses daunting problems. The ECVAM studies summarized previously were exceptional in that the authors took the trouble to try and verify and standardize the readout of the assay that is inherently subjective and lacks the necessary external standards and controls by which the assay could be properly validated. In studies performed by the National Marrow Donor Program (NMDP), the results showed very high variability in CFU colony counting for cord blood [33]. This high variability, primarily due to the inaccuracy of dispensing methylcellulose and colony counting, together with the lack of high throughput capability does not provide the biopharmaceutical industry, environmental agencies or other areas of toxicology, risk or efficacy assessment with a routine and trustworthy assay platform. To negate all of these problems, the HALO Predictive Hemotoxicity Platform was developed.

4. Predictive stem cell hemotoxicity testing

Whereas the CFU assay may be used to predict neutropenia, thrombocytopenia, anemia and the MTD indicated by the IC90 values [27-29], stem cells assays allow potential hemotoxicity to be taken to a different system-wide "global" level. The reason is provided in Fig. 2 and in more detail in Fig. 4, which shows the different lympho-hematopoietic cell populations that can be detected using a hemotoxicity screening and testing platform specifically developed for this purpose. This platform, called HALO, will be described in more detail in the next section. Figures 2 and 4 demonstrate that functionally mature cells from definitive continuously proliferating and partially proliferating cell systems, are derived from stem cells. As such, any perturbation or damage to the stem cell compartment will ultimately affect all downstream cell populations. In other words, examining the effect on stem cells allows the "global" effect on the system to be predicted. Since more is known about the organization, hierarchy and regulation of the lympho-hematopoietic system than probably any other biological system in the body, this knowledge can be used to predict and explain potentially deleterious effects to the system. Changes in the response to hematopoietic stem cells will affect all three primary hematopoietic lineages, namely the erythropoietic, myelomonocytic and magakaryopoietic lineages. Changes in the response to lympho-hematopoietic stem cells, i.e. those stem cells that can give rise to both the lymphopoietic and hematopoietic cells, will be expected to affect most, if not all cell lineages, including the T- and B-cell lineages and therefore the immune system as a whole.

Predictive stem cell hemotoxicity testing is not simply the estimation of IC values so that compounds can be ranked in order of toxicity to different cell populations or species. There are several other important applications in which stem cell hemotoxicity, and indeed stem cell toxicity in general, can be used. Examples of these applications will be discussed later in this chapter. First, however, it is necessary to describe the principles, characteristics and properties of the assay that make this possible.

5. Materials and methods

HALO is the acronym for Hematopoietic/Hemotoxicity Assays via Luminescence Output. This platform was originally designed and developed to provide the biopharmaceutical industry with a high throughput, validated assay to examine the effects of virtually any compound on different cell populations of the lympho-hematopoietic system from multiple species. Initially, the assay platform was developed for fresh, primary human cells, as a surrogate assay that could be used at virtually at stage in the drug development pipeline (Fig. 3) to extrapolate to the human situation, and as an alternative to pre-clinical animal studies. The platform has since been further developed to include non-human primate, horse, pig, sheep, dog, rat and mouse, not only for toxicity studies, but also for basic research and veterinary applications.

http://dx.doi.org/10.5772/54430

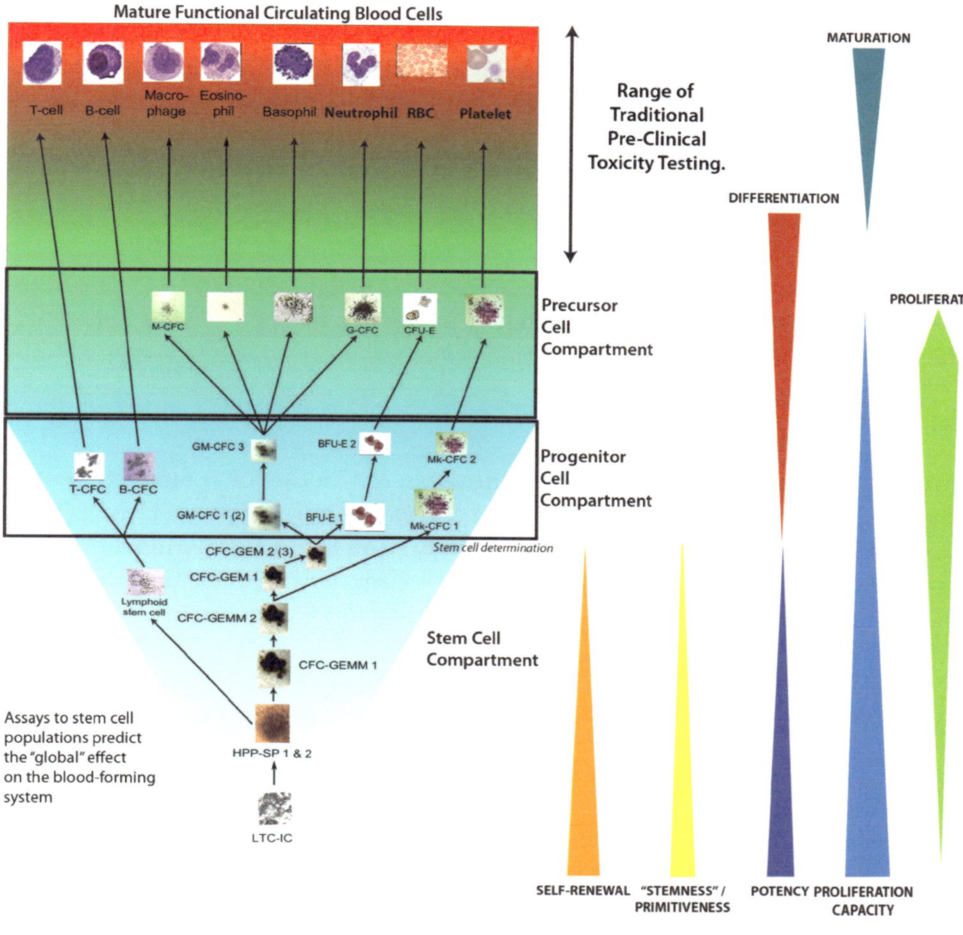

Figure 4. The Organization and Hierarchy of the Lympho-Hematopoietic System as a Model for a Definitive Continuously Proliferating Stem Cell System. The properties of stem cells play an integral part in predicting toxicity.

5.1. Concepts and principles of the HALO platform

When cells proliferate or are inhibited from proliferation by drugs or other agents, the concentration of intracellular adenosine triphosphate (iATP) changes proportionately. This biochemical marker is an indicator of cellular and mitochondrial integrity and therefore viability of the cells. Indeed, iATP is used as a metabolic viability assay (as opposed to a dye exclusion viability assay). Under normal conditions, stimulation of cell proliferation requires specific growth factors and/or cytokines either alone or in combination (cocktails). For continuously proliferating systems, growth factors or cytokines need to be present continuously, albeit, in very small concentrations, in order to maintain cell survival and production. Thus, to detect the effect of any agent on hematopoietic cells *in vitro*, the target cell population must be stimulated in order to detect changes in the cell population response to the agent. The agent is usually added in a dose-dependent manner to the target cells, which are then incubated for a specific period of time. Thereafter, the cultures are removed from the incubator and the cells lysed to release the iATP. The latter then becomes a limiting substrate for a luciferin-luciferase reaction to produce bioluminescence in the form of light as shown in the equation below.

$$i\text{ATP} + \text{Luciferin} + O_2 \xrightarrow[Mg^{2+}]{\text{Luciferase}} \text{Oxyluciferin} + \text{AMP} + \text{PP}_i + CO_2 + \textbf{\textit{LIGHT}}$$

The light is measured in a plate luminometer. The amount of light produced correlates directly with any change in the iATP concentration and therefore with the state of proliferation or inhibition of the cells.

5.2. Cell sources

Cells from any hematopoietic or lymphopoietic organ can be used. For most of the studies described here, fresh or cryopreserved human bone marrow or peripheral blood cells were collected with prior authorization by an Internal Review Board (IRB). Human cells were obtained from Lonza (Walkerville, MD) or Allcells (Berkely, CA). A mononuclear cell (MNC) fraction was prepared using density gradient centrifugation. A nucleated cell count was performed using a Z2 particle counter (Beckman Coulter), while dye exclusion viability was performed using 7-aminoactinoycin D (7-AAD) and flow cytometry. Metabolic viability was performed using LIVEGlo (HemoGenix, Colorado Springs, CO).

5.3. Cell culture

For all toxicity studies, MNC were diluted so that the final cell concentration was either 7,500 or 10,000 cells/well. Either 96-well or 384-well, solid white-wall plates were used and all dispensing was performed using a liquid handler (Beckman Coulter, EPICS XL-MCL). After the cell suspension was prepared, it was added to a Master Mix containing reagents including growth factors and/or cytokines to stimulate the target cell population being studied. Five different hematopoietic stem cell populations have so far been developed for this assay, the most important being the Colony-Forming Cell – Granulocyte, Erythroid, Macrophage, Megakaryocyte or CFC-GEMM (referred to in Fig. 4 as CFC-GEMM 1). This particular stem cell population is stimulated with erythropoietin (EPO), granulocyte-macrophage and granulocyte colony stimulating factors (GM-CSF, G-CSF), Interleukins 3 and 6 (IL-3, IL-6), stem cell factor (SCF), Flt3-Ligand (Flt3-L) and thrombopoietin (TPO). Compared to a "classic" CFC assay, HALO does not incorporate methylcellulose and is therefore not a clonal assay. Instead HALO uses Suspension Expansion Culture (SEC) Technology, which has several advantages over methylcellulose assays. First, SEC assays allow more accurate dispensing using liquid handlers. This is in contrast to inaccurately dispensing methylcellulose with syringes and needles. Second, the use of liquid handlers allows for true high throughput capability with accurate dispensing even in 384-well plates. Third, as opposed to methylcellulose, where little or no cell interaction occurs, SEC technology allows cells to interact with each other. This has two important consequences. Cell interaction reduces the time for the onset of cell proliferation by approximately 24 hours. This means that measurement of cell proliferation can be measured within 5 to 7 days. Indeed, for all of the studies described here, human cells were incubated for 5 days. Non-human primate cells are usually incubated for the same time, but all other animal cells only require 4 days of incubation. The second consequence of allowing cell interaction to occur is the two-fold increase in assay sensitivity. As with most cell cultures, cells are incubated at 37ºC in a fully humidified atmosphere containing CO_2. Incubating cells under low oxygen tension of 5% O_2, which is approx. equivalent to the venous oxygen tension, reduces oxygen toxicity due to free radical production and improves plating efficiency [34,35] for all lympho-hematopoietic cell populations as well as other cell types.

5.4. Controls and dosing

Four basic controls were always included for toxicity studies. A background control included cells, but no growth factors. A vehicle control was similar to the background control, but included the vehicle used to dissolve the compound. Growth of the target cell population without any compound or vehicle constituted the growth factor control. A similar control that included the vehicle was designated the growth factor + vehicle control. Drugs and other agents were investigated over 6 – 9 doses.

5.5. Instrument calibration, assay standardization and sample processing

Prior to measuring any sample, the instrument was calibrated and the assay standardized using an external ATP standard and controls. The procedures have been described previously [32] and detailed procedures can also be obtained [36,37]. Calibration and standardization were also part of the assay validation process (see Section 5.6).

There are other advantages for calibrating and standardizing the assay. First results can be compared over time. Second, the output of a plate luminometer is in Relative Luminescence Units or RLU. The results are relative because different instruments demonstrate different ranges of RLU. These ranges may vary from 0 to 100 for one manufacturer or 0 to several million for another. This means that it would be very difficult to directly compare results within and between laboratories using RLU values. Performing an ATP standard curve allows all the results to be interpolated from RLU values into standardized ATP concentrations (μM).

5.6. Assay verification and validation

HALO was originally developed from the "classic" CFC assay because the latter was the only cell-based assay that could detect primitive hematopoietic cell populations. Since HALO is a proliferation assay, while the CFC detects differentiation of the same cells, and because proliferation occurs prior to differentiation, it follows that one assay can verify the other. Indeed, several publications have shown a direct correlation between the two assays [32,38,39].

Validation, on the other hand, is quite a different matter. Assay validation is defined as "establishing documented evidence which provides a high degree of assurance that a specific process will consistently produce a product meeting its predetermined specifications and quality attributes" [40]. When an assay is properly validated the accuracy (proportion of correct outcomes), sensitivity (proportion of correctly identified positive samples), selectivity (proportion of correctly identified negative samples), precision (intra and inter-laboratory variability) and robustness (the ability of the assay to withstand changes and transferability) all combine to give the user the assurance that the results obtained are correct. The ECVAM studies described in Section 2 above were, and still are, the closest the CFC assay has come to being validated. There have been many attempts to validate the CFC assay, but all have failed. Certainly the assay has shown, from a subjective viewpoint, some of the attributes. However, since there are no standards and controls by which the CFC assay can provide documented and quantitative evidence for each of the required parameters, the assay has never been properly validated. Like many assays that have been used for decades, the CFC assay has been "grandfathered" in and used despite the problematic trustworthiness and meaning of the results obtained [33,37].

HALO, from the outset, was designed to be validated. The assay was developed to incorporate the range values specified in the FDA Guidance on Bioanalytical Method Validation [40]. In summary, these values are as follows:

- Assay linearity: => 5 logs.
- Assay cell linearity: 1,000 - > 25,000 cells/well.
- Assay ATP sensitivity: ~ 0.001μM.
- Assay cell sensitivity: 20-25 cells/well, depending on cell purity).
- Accuracy: ~95%.
- Sensitivity & Selectivity by Receiver Operator Characteristics (ROC): Area Under Curve (AUC) 0.73 – 0.752 (lowest possible value: 0.5; highest possible value, 1).
- Precision: = < 15%. Lower limit of quantification (LLOQ): 20%.
- Robustness: ~95%.
- High throughput capability (Z-factor [57]): > 0.76.
- Log-log linear regression slope for ATP standard curve: 0.937 ± 15% (slope range: 0.796 – 1.07)
- Lowest ATP value indicating unsustainable cell proliferation: ~ 0.04μM.
- ATP value below which cells are not metabolically viable: ~0.01μM.

In addition, the assay has also been validated against the Registry of Cytotoxicity Prediction Model, which will be discussed in more detail in Section 5B.

5.7. Statistics

All of the results provided were produced using 8 replicate wells/point. Compound dose response curves were fitted to a 4- or 5-parameter logistic curve fit using SoftMax Pro software (Molecular Devices) from results exported directed from the plate luminometer and calculated automatically. To estimate IC values, raw data were converted to a percentage of the growth factor + vehicle control. Additional statistics, curve fitting or graphing was performed using Prism software (GraphPad) or OriginPro (OriginLab).

6. Results and discussion

6.1. Distinguishing the response of stem cells from progenitor cells

From a practical viewpoint, stem and progenitor cells are distinguished by at least two different characteristics. First, stem and progenitor cell populations are stimulated using different cocktails of growth factors and cytokines. In this way, specific cell populations can be targeted and studied, even though the cell suspension may contain other cell types. Combined with the culture conditions, this allows detection and measurement of specific cell populations. The other distinguishing characteristic is the difference in proliferation ability and potential between stem and progenitor cells. Even within the stem cell compartment, differences in proliferation potential will indicate the primitiveness or "stemness" of populations. This characteristic is shown in Fig. 5 for normal bone marrow cells. Since the stem cells are more primitive than the progenitor cells, it would be expected that their proliferation potential would be greater. Figure 5 shows that the two stem cell populations exhibit, not only greater ATP concentration values, but also greater linear regression cell dose response slopes than the hematopoietic or lymphopoietic progenitor cells. It is the slope of the cell dose response that measures proliferation potential. The greater the slope, the higher the proliferation potential, and the more primitive the cell population. Indeed, this is the basic principle for measuring potency of hematopoietic stem cell therapeutic products for transplantation [37]. In this way, it is possible to distinguish different stem cell populations, in this case the hematopoietic stem cell, CFC-GEMM 1, from the more primitive lympho-hematopoietic stem cell, HPP-SP (high proliferative potential – stem and progenitor cell). The HPP-SP stem cell will be discussed in more detail in Section 6.4. The three cell dose response clusters showing the differences in proliferation potential in Fig. 5 for stem cells, hematopoietic progenitor cells and lymphopoietic progenitor cells would be expected based on the organization of the blood-forming system shown in Fig. 4. Figure 6 demonstrates the expected proliferation ability of the seven different cell populations in response to mitomycin-C, with the stem cells showing the greatest ability to proliferate followed by the three hematopoietic lineages and lymphopoietic lineages.

The steepness of the linear regression slope of the cell dose response for a cell population provides a measure of the proliferation potential. Stem cells exhibit the greatest proliferation potential of all cells. Within the stem cell compartment, stem cells with different potentials for proliferation also indicate their primitiveness. Proliferation ability is measured at a single cell dose (see Fig. 6).

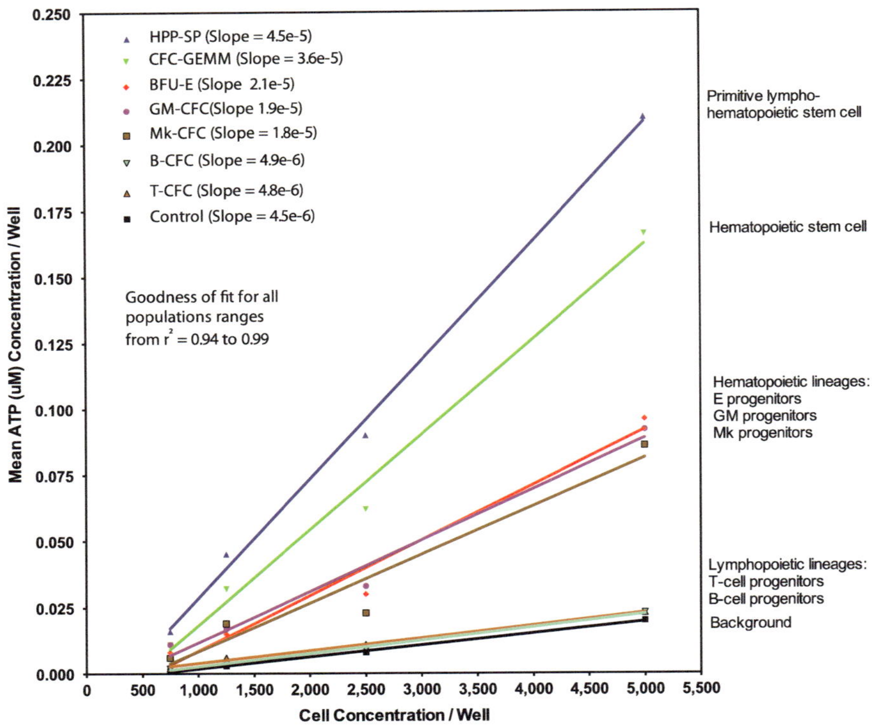

Figure 5. Measuring Proliferation Potential of Cell Populations

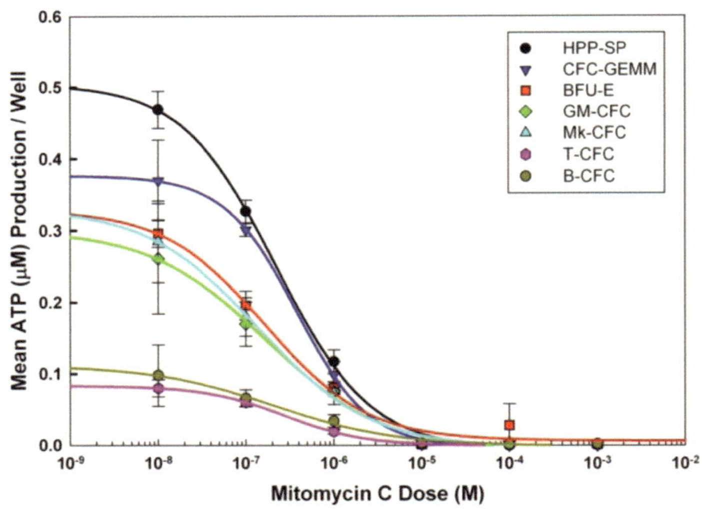

Figure 6. Demonstration of Proliferation Ability between Cell Populations

6.2. Drug and compound screening for stem cell toxicity

In its most basic form, a single drug or compound is tested in a dose dependent manner on a target cell population. If the agent is cytotoxic to the cells, a negative sigmoidal dose response (Fig. 6) will result from which the estimated percent inhibitory concentrations (IC) can be calculated. Figure 7 shows the dose response curves from 13 drugs and compounds tested on hematopoietic stem cells (CFC-GEMM 1) derived from fresh, human bone marrow using the MNC fraction. Although different cell types are included in this fraction, stimulation of this particular stem cell population using a specific growth factor cocktail provides the relevant information. For each of the compounds tested a 4-parameter logistic curve fit was plotted from which the IC values could be calculated. Table 1 shows all of the compounds ranked in order of IC50 (μM) value from the most to the least toxic. The IC90 value (equivalent to the maximum tolerated dose, MTD) is also provided. Many of the compounds tested were also used in the ECVAM studies [26,27].

Table 1 shows some compounds designated as NV or NE. The term NV indicates that an IC20 values was obtained, but no IC50 or IC90 value. The term NE means "no effect" in that no IC values could be estimated. As a result, methotrexate, which is an anti-cancer agent and expected to produce a more dramatic effect on stem cells, is actually ranked near the end of the list. Furthermore, compounds that do not allow an IC value to be calculated might actually produce some effect. The problem with ranking compounds based on their IC values is that it does not take into account the "form" of the dose response curve, which can actually provide more information than the IC value alone. Figure 7 shows a large number of different dose response curves. One of the most important parameters provided by the 4-parameter logistic curve fit is coefficient or parameter B, which describes the transition of the curve to the midpoint of the dose response. This is a measure of steepness or slope. In some cases the slope is shallow, while in other cases it is almost vertical. How can this and other parameters of the dose response curve be taken into account so that they are independent of the IC value? The answer lies in calculating the area under the curve (AUC) for the range of doses used. When the AUC is performed and plotted so that the compounds are ranked, a different and more plausible picture is obtained (Fig. 8).

In this case, the AUC values for both stem cells (CFC-GEMM 1) and granulocyte-macrophage colony-forming cells (GM-CFC) are shown. When the results for CFC-GEMM 1 are compared with those in Table 1, the results generally follow the IC50 values. However, the toxicity of methotrexate is significantly increased and cycloheximide is more toxic than paclitaxel. The results for the GM progenitor cells have been included to demonstrate that progenitor cells exhibit lower toxicities than stem cells. Unless there is evidence to demonstrate that a compound acts on a specific hematopoietic lineage, it is more prudent to analyze potential toxicity to the stem cell compartment first, rather than focusing on a particular lineage, since the latter will only provide limited information that could possibly result in a false interpretation and conclusion.

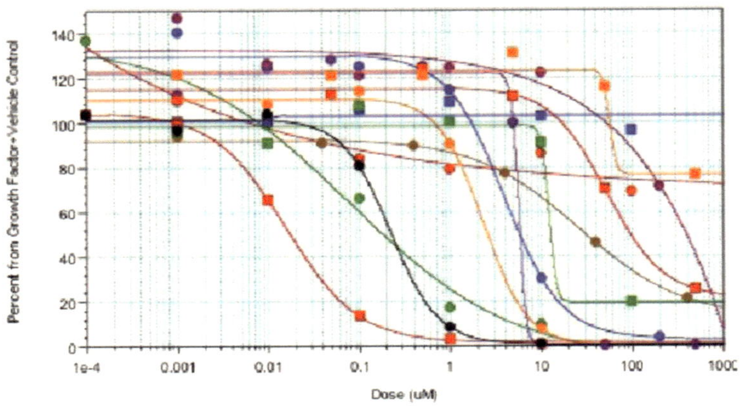

4-P Fit: y = (A - D)/(1 + (x/C)^B) + D:	A	B	C	D	R^2
5-Flurouracil (C1-CFC-GEMM: Conc (uM) vs Mean%...	130	1.39	3.97	2.72	0.991
Methotrexate (C2-CFC-GEMM: Conc (uM) vs Mean%...	5.11e+08	0.18	4.48e-43	68.5	0.881
Paclitaxel (C3-CFC-GEMM: Conc (uM) vs Mean%GF+...	136	0.477	0.0795	-7.79	0.979
Cyclosporin A (C4-CFC-GEMM: Conc (uM) vs Mean...	110	1.8	2.32	-0.173	0.999
Indomethacin (C5-CFC-GEMM: Conc (uM) vs Mean%...	132	0.451	2.59e+14	-1.75e+07	0.882
AZT (C6-CFC-GEMM: Conc (uM) vs Mean%GF+Vhl)	92.3	0.768	26.4	12	0.999
Chlorpromazine (C7-CFC-GEMM: Conc (uM) vs Mean...	122	11.8	5.66	0.0478	0.993
Acyclovir (C8-CFC-GEMM: Conc (uM) vs Mean%GF...	100	12.4	0.0108	103	0.0847
Camptothecin (C9-CFC-GEMM: Conc (uM) vs Mean%...	104	1.1	0.0159	1.17	1
Choramphenicol (C10-CFC-GEMM: Conc (uM) vs Me...	115	1.41	53.1	21.2	0.984
Rifampicin (C11-CFC-GEMM: Conc (uM) vs Mean%G...	98.3	11.3	12.2	19.2	0.973
Warfarin (C12-CFC-GEMM: Conc (uM) vs Mean%GF...	123	12	57	76.2	0.96
Cycloheximide (C13-CFC-GEMM: Conc (uM) vs Mean...	101	1.67	0.225	0.41	0.997

Figure 7. The Effect of 13 Compounds on Hematopoietic CFC-GEMM Stem Cells. Diagram showing the dose response plots produced automatically by SoftMax Pro software after the data was collected by the SpectraMax L plate luminometer. The parameters that define the 4-parameter logistic curve to which the dose responses of the compounds are fitted are as follows: Parameter A, asymptote (flat part of the curve) at low Y-values; Parameter D, asymptote at the highest Y-values; Parameter or coefficient B, the transition from the asymptotes to the center of the curve; Parameter or coefficient C, is the midpoint between parameters A and D, also called the IC50 or EC50. Data that cannot be properly fitted will result in ambiguous results.

Compound	Effect	Rank	IC50 (μM)	IC90 (μM)
Camptothecin	Anti-cancer	1	0.02	0.14
Paclitaxol	Anti-cancer	2	0.18	4.81
Cycloheximide	Pesticide	3	0.23	0.86
Cyclosporin A	Immunosuppressant	4	2.57	8.2
5-Fluorouracil	Anti-cancer	5	5.79	29.7
Chlorpromazine (Thorazine)	Anti-psychotic	6	5.88	7.03
Rifampicin	Anti-bacterial	7	12.8	NV
Zedovuidine(AZT)	Anti-viral	8	30.4	NV
Choramphenicol	Anti-bacterial	9	94.7	NV
Indomethacin	Anti-inflammatory	10	394.2	947.5
Methotrexate	Anti-cancer	11	NV	NV
Acyclovir	Anti-viral	12	NE	NE
Warfarin	Anti-coagulant	13	NE	NE

NV indicates No Value for these IC values. An IC20 value would have been estimated by the software program.

NE indicate No Effect. In this case, the dose response for the compound did not produce an IC values.

Table 1. Ranking of Stem Cell Toxicity According to IC50 Values

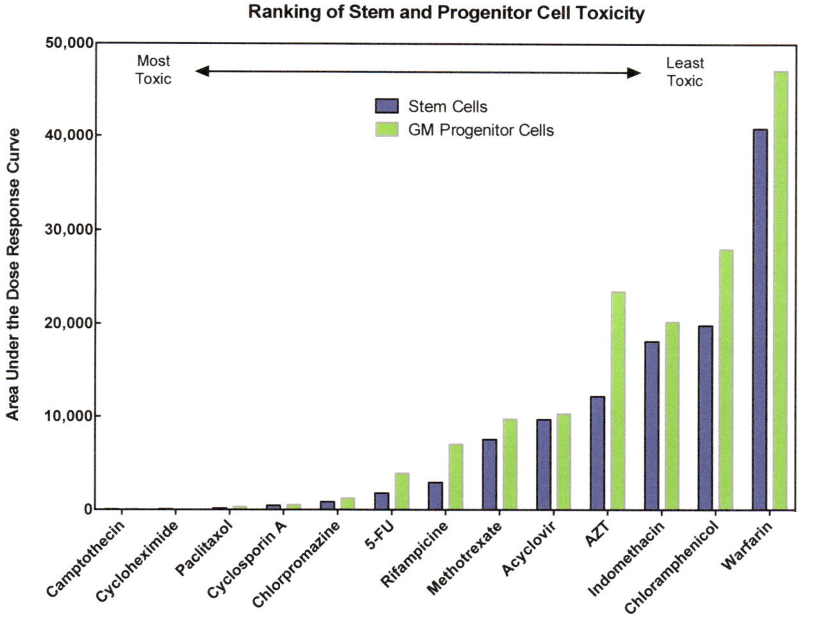

Figure 8. Ranking of Stem Cell and Granulocyte-Macrophage Progenitor (GM-CFC) Toxicity According to the Calculated Area Under the Curve (AUC) for the Dose Responses shown in Figure 7.

6.3. The registry of cytotoxicity prediction model [41]

The Registry of Cytotoxicity (RC) is a list of 347 compounds, for which the IC50 values using a neutral red uptake assay for human keratinocytes and mouse 3T3 cells and the oral LD50 values for rat or mouse, are known. When validating an *in vitro* assay against the RC, a sample of reference compounds is tested. The resulting IC50 values from the *in vitro* assay are then plotted against the LD50 values for the same compounds. A linear regression should be obtained exhibiting equation constants within a specific range. If this occurs, the *in vitro* assay is considered a validated cytotoxic test. The results validating HALO against the RC Prediction Model were first reported in 2005 [33]. One of the most interesting aspects of this prediction model is that once an assay has been validated, it can be used to convert *in vitro* IC values into clinically relevant doses that can be used as starting doses for pre-clinical animal models or human clinical trials. An example of this is shown in Table 2 where the results of converting IC20, IC50 and IC90 range values derived from the effects of 18 compounds on CFC-GEMM 1 bone marrow cells is shown. The predicted doses derived from the IC values are given in both milligrams/kilogram (mg/kg) and milligrams/meter^2 (mg/m^2). Doses used in the clinic to treat patients are also shown in mg/kg or mg/m^2 where available. With the exception of two drugs, namely acyclovir and warfarin, nearly all of the doses predicted by the *in vitro* CFC-GEMM 1 assay using ATP bioluminescence are in the same order of magnitude or very close to the doses used to treat patients. In some cases lower doses were predicted (e.g. 5-fluorouracil), while in other cases slightly higher doses were predicted (e.g. cyclosporine A, indomethacin, cisplatin and mitomycin-C). Thus these predicted starting values may be used in early toxicity and efficacy studies to "bracket" the lower and higher dose ranges.

6.4. Residual stem cells after toxicity

Figure 7 shows that the response of stem cells to toxic agents can vary dramatically. In some cases, agents cause complete eradication of all stem and progenitor cells at high doses. In other cases, there is partial cytotoxicity at which, even at high doses, stem and progenitor cells are not eradicated. This is an indication that some stem cells survive or are possibly resistant to the drug or compound. If stem cells are not noticeably affected at high doses, there is a good chance that when the drug or compound is removed, the system will reconstitute itself. If no stem cells are available, this will not occur. However, there are other aspects to this phenomenon that are important.

Primitive stem cells are usually in a quiescent state; they are not proliferating and therefore not in cell cycle. This does not mean that they cannot be affected by an agent. Small molecules can enter a cell even if it is quiescent. When required to initiate the proliferation process and begin cell division, the process may be aborted because the agent inhibits the process. This is a potential dangerous situation for two reasons. First, the "backup plan" for reconstituting the system may not function. Second, if cells do begin to proliferate and divide, they may be more sensitive to the agent. The consequence of this is that repeated administration of the drug or compound will continually reduce the proportion of residual stem cells present.

	Predicted Dosing Range from In Vitro Stem Cell Assay		**Published Drug Doses Used to Treat Patients**	
Drug/Compound	Dose in mg/kg	Dose in mg/m^2	Doses or Dose Range in mg/kg	Doses or Dose Range in mg/m^2
Doxorubicin	2.6 – 6.9	97 – 255		25/50/60/75
Daunorubicin	0.5 – 2.6	19.6 – 97		30/45/60
5-Fluorouracil	2.0 – 7.0	79 – 259		400 – 2,600
Paclitaxel	2.0 – 17.5	72 – 647		75 – 250
Imatinib (Gleevec)	3.6 – 30.5	132 – 1,125		400/600
Methotrexate	5.8	215		10 – 8,000
Cyclosporin A	14.2 – 31.2	524 – 1,155	5 – 10	
Indomethacin	32 – 73	1,190 – 2,700	0.2 – 2	
Zedovudine (AZT)	4.3 – 12.2	161 – 452	1 – 7.4	
Chlorpromozine (Thorazine)	6.8 – 7.7	253 – 285	1 – 4.5	
Acyclovir	NV	NV	5 – 500	
Camptothecin	0.36 – 1.52	13.3 – 56		25/320/470
Choramphenicol	16 – 24	594 – 896	12.5/30 – 50	
Rifampicin	24 – 26	894 – 955	10	
Warfarin	NV	NV	0.1 – 5	
SJG-136	0.1 – 0.3	4 – 10		6 – 40
Cisplatin	6.3 – 9.8	233 – 363		30 – 100
Mitomycin-C	1.2 – 6.0	47 – 220		6/10 – 20

The IC values obtained from the validated *in vitro* assay are entered into the equation: Y = 0.435 * Log (IC value) + 0.625 [41]. The dose in mg/kg is then obtained by multiplying the value for Y with the molecular weight of the compound. The dose in mg/m^2 is obtained by multiplying the dose in mg/kg by a specific factor described in [42].

Table 2. Using the Registry of Cytotoxicity Prediction Model to Convert *In Vitro* IC Values into Clinically Relevant Starting Doses

To demonstrate this, we developed an *in vitro* secondary re-plating assay for primitive stem cells called high proliferation potential – stem and progenitor cells (HPP-SP). This stem cell population, within the stem cell compartment (Fig. 4), is approximately at the divergence of the lymphopoietic and hematopoietic systems. The majority of HPP-SP stem cells are quiescent. They can be induced or "primed" into proliferation with IL-3, IL-6, SCF and Fl3-L. This stem cell population is designated HPP-SP 1. Once the HPP-SP 1 cells begin proliferation, they can be expanded with a similar cocktail of growth factors and cytokines to that for CFC-GEMM 1, but with the addition of interleukins 2 and 7 (IL-2, IL-7). This fully stimulated primitive stem cell population is designated HPP-SP 2. In this two-stage assay, the HPP-SP 1, present in the MNC fraction of bone marrow are cultured in the presence of the drug or compound in a dose-dependent manner. Thereafter, the cells are removed from culture, washed and re-plated in a secondary culture system in which the HPP-SP 2 population is measured. By performing a secondary re-plating step, the assay is substantiating the presence of primitive stem cells present in the first "priming" step of culture. The proliferation at both stages is determined using ATP bioluminescence technology. The results using busulphan and daunorubicin are shown in Figs. 9A and 9B, respectively. The effect of busulphan (Fig. 9A) on HPP-SP 1 demonstrates partial cytotoxicity to the stem cells and the presence of residual stem cells. However, when the treated cells are removed from primary culture and placed into secondary cultured to reveal their expansion potential, there are few residual cells that are available for expansion and the high doses used in the primary culture eradicated any remaining cells. There was also little change in the IC50 values. This indicates that busulphan continued to act on primitive stem cells leaving no residual stem cells (secondary culture results minus primary culture results) for possible repopulation. Daunorubicin (Fig. 9B) is highly toxic to stem cells with an IC50 value in the nanomolar range compared to the micromolar range for busulphan. At low doses of daunorubicin, residual stem cells would be available, but secondary culture demonstrates that both these and the residual cells have increased their sensitivity by approx. 3 fold, indicating that repeated drug administration would incur increased sensitivity of the stem cells to the drug.

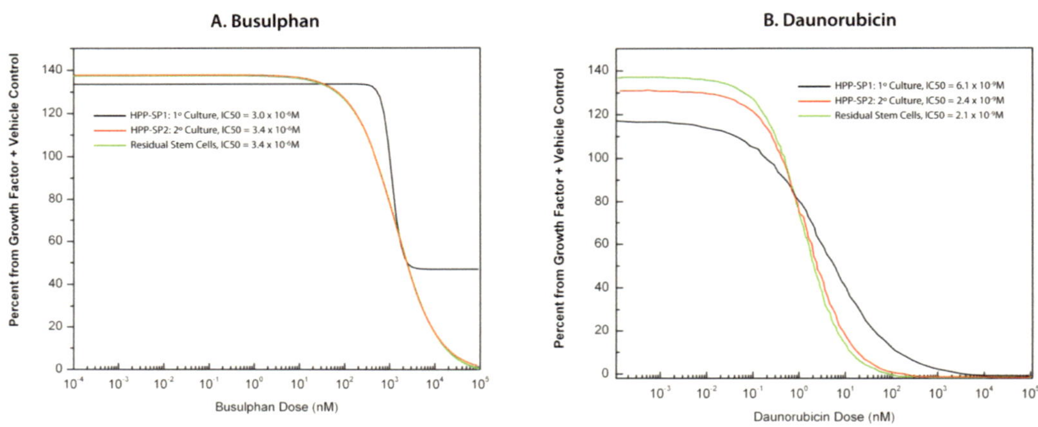

Figure 9. Assessing Residual Stem Cell Activity and Change in Stem Cell Sensitivity to Agents by Measuring the Response of Primitive Stem Cells in a Two-Step Secondary Re-Plating In Vitro Assay.

6.5. Stem cells and drug-drug interactions

Drug-drug interaction (DDI) can lead to dangerous consequences if not investigated properly. Traditionally, DDI are investigated using cultured hepatocytes since the liver is the organ primarily responsible for detoxification. The main enzymes investigated during DDI studies are those of the cytochrome P450 (CYP450) system present in the endoplasmic reticulum of the cells. CYP450 enzymes are present not only in hepatocytes, but in virtually all cells. There are a large number of CYP450 enzymes and assays are available for many of these. Depending on the drug or compound, one or more CYP450 enzymes can be induced or inhibited [43,44]. The response by different enzymes provides an indication as to whether an interaction between different drugs will occur. However, measurement of CYP450 activities does not indicate a response at the cellular level. To investigate this, we developed an assay in which drugs could be titrated against each other to determine potential DDI on stem cells.

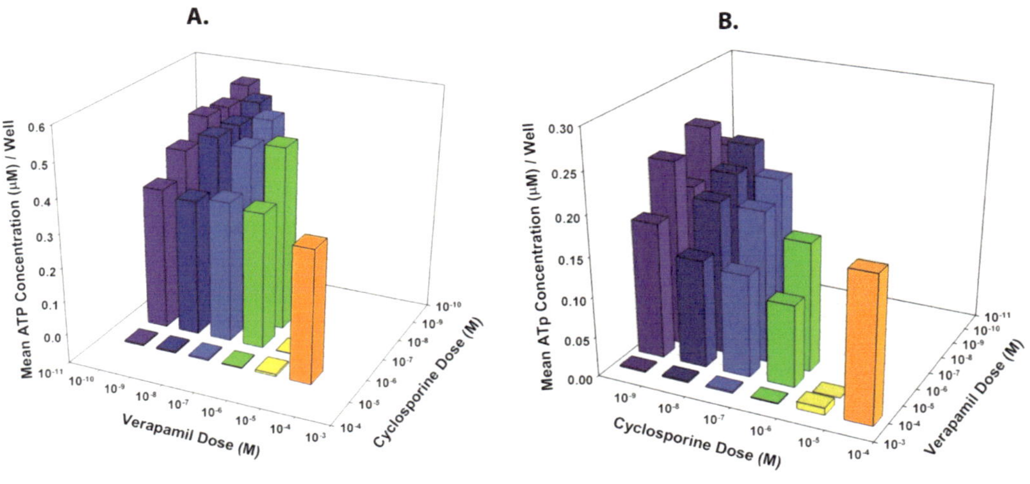

Figure 10. Examples of Drug-Drug Interactions at the Stem Cell Level

Figure 10A shows the response when verapamil is titrated against cyclosporin A, while Fig. 10B shows the effect when cyclosporin A is titrated against verapamil. Both drugs inhibit 3A4 CYP450 enzyme. Individually, both drugs are cytotoxic to CFC-GEMM 1 stem cells. However, when titrated against each other, cytotoxicity may be observed initially, but may be followed by an opposite effect at higher doses. The cells appear to overcome the inhibitory effects. In terms of DDI, this would indicate that one or other drug is present at concentrations that could cause serious harm to the patient. This unusual dose response behavior produces a U-shaped or inverted dose response curve that has been observed for many compounds, including dopamine [45] and endostatins [46]. Although often attributed to solubility, these effects appear to be pharmacologically and physiologically important, but in most cases, the mechanism is not understood. This is the first indication that DDI can occur at the stem cell level. Considering the importance of assessing toxicity to stem cells and the predictive value afforded by these cells, it is obvious that more has to be learnt before the consequences of these reactions on a stem cell system can be understood.

6.6. Circadian rhythm and stem cells

One of the most interesting aspects of drug treatment is the field of chronotherapy; the administration of drugs in accordance with circadian rhythms. Although studied for decades, the role of circadian rhythms to reduce toxicity and improve drug efficacy has been largely ignored by the biopharmaceutical industry. The primary reason for this is because chronotherapeutic studies are difficult, time-consuming and expensive to perform. Nevertheless, many areas of chronotherapy, especially using anti-cancer drugs. have proved to be successful [47-49]. Many cellular functions are dependent upon circadian rhythms. It is not the purpose of this section to describe or even summarize this field. The intention is to instead provide an example in which the circadian rhythm of cells, especially hematopoietic stem cells [50-52], can be used to predict the best time of day to administer an anti-cancer drug, which in this case, is 5-fluorouracil (5-FU) [38].

These studies were performed using normal peripheral blood mononuclear cells. Blood was obtained from the same donor every 4 hours over a 24 hours period. The MNCs were fractionated at each time point and cryopreserved into aliquots. Prior to cryopreservation, an aliquot of fresh cells was used to measure the proliferation ability of hematopoietic stem cells (CFC-GEMM 1), erythropoietic progenitor cells (burst-forming units – erythroid, BFU-E), GM-CFC and megakaryopoietic progenitor cells (megakaryopoietic colony-forming cells, Mk-CFC) at each time point using HALO. After collection of the cells, an aliquot from each time point was thawed and the circadian rhythms compared to fresh cells. A cosinor curve fitting analysis was performed to produce all the circadian rhythms shown in Fig. 11 [53]. The results for hematopoietic stem cells (Fig. 11A) and all progenitor cells (not shown) demonstrate that even after cryopreservation, the cell populations maintain their circadian rhythm. This was a prerequisite to use cryopreserved cells for the remainder of the study.

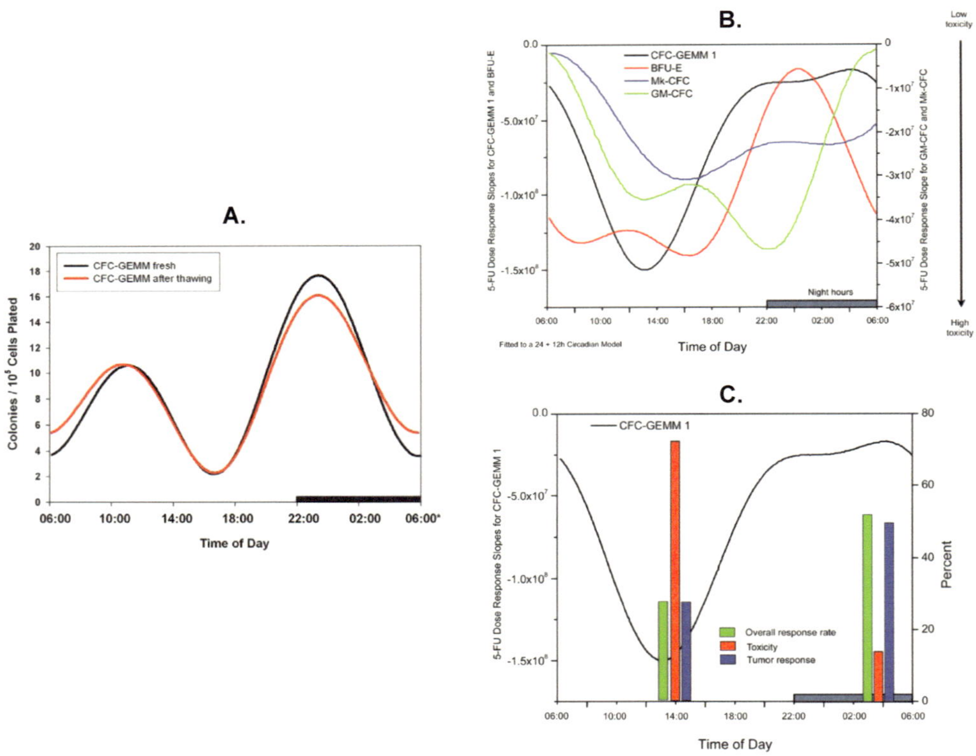

Figure 11. Using the Circadian Rhythm of Hematopoietic Stem Cells to Predict the Best Time of Day to Administer 5-Fluorouracil to Reduce Toxicity and Improve Efficacy of the Drug.

For each time point, cells were thawed and treated with 5-FU at six doses to measure the response of CFC-GEMM 1, BFU-E, GM-CFC and Mk-CFC. The slope of each negative sigmoidal dose response curve was then calculated from the 4-parameter logistic curve fit. The dose response slope values were then analyzed by cosinor analysis for each time point and for each cell population to obtain the circadian rhythms as a function of 5-FU treatment. The results are shown in Fig. 10B. Each of the hematopoietic cell populations exhibited its own circadian rhythm in response to 5-FU. When these circadian rhythms were correlated with either the continuous infusion of 5-FU that is normally used to treat patients and that of chronomodulated infusion of 5-FU as reported by Dogliotti and colleagues in 1998 [54], the results shown in Fig. 10C were obtained. For each of the administration types, the percent overall patient response rate, toxicity and tumor response are shown. These were overlaid onto the circadian rhythm for the CFC-GEMM 1 stem cell response to 5-FU and demonstrated that the lowest toxicity and highest overall and tumor response occurred when 5-FU was administered in a chronomodulated manner in the early morning hours rather than at any other time of the day. The nadir of the CFC-GEMM 1 circadian rhythm to 5-FU occurred at 14:00 hours in the afternoon. This was approximately the same time at which the highest toxicity to 5-FU was found. As expected, these results did not correlate nearly as well for the hematopoietic progenitor cells. In addition, the results clearly demonstrate that the potential for toxicity can be dramatically reduced if the circadian rhythm of the target cells is taken into account. From the brief description here, it follows that to ascertain the best time of day to administer a drug a considerable amount of work must be undertaken. The question is whether the patient response and well-being outweigh the time and cost to perform these types of studies.

7. Conclusions and future trends

To use *in vitro* stem cell assays to predict potential toxicity to the hematopoietic system, and any stem cell system for that matter, knowledge of the biology, physiology, regulation and response is required for an *in vitro* to *in vivo* concordance to be justified. This concordance plays an integral role in predicting toxicity since it allows for *in vitro* surrogate assays to be used in place of animals and therefore comply with the principle of the 3Rs (replacement, reduction and refinement) [55]. More importantly, to allows extrapolation to the human situation. Previous literature on stem cell and hematopoietic research demonstrates that *in vitro* assays show a high concordance with *in vivo* data. Using the HALO platform, Olaharski et al. demonstrated an *in vitro* to *in vivo* concordance of greater than 80% [56]. This high degree of concordance provides the basis to predict the response of the lympho-hematopoietic and other stem cell systems to potential toxic insults. This has been described previously [32], but it is worth reiterating some of these paradigms. First, virtually any compound can be toxic to stem cells. Second, toxicity to the most primitive, definitive stem cells will affect all cells of the system. Third, since stem cells only proliferate and proliferation occurs prior to differentiation, stem cell cytotoxicity will affect all downstream cell types. Fourth, if more than one cell lineage is affected by toxicity, the target is not the cells that constitute the lineages, but the stem cells producing the lineages. Finally, stem cells are more sensitive to toxicity than the progenitor cells. When considering using stem cells to predict potential toxicity, at least two considerations need to be taken into account. The first is the primitiveness of the stem cell population being measured, while the second is variation between human donors. The former will depend, among other things, upon the ability and sensitivity of the assay to detect specific stem cell populations and the latter will be dependent upon the state and demographics of the donors that can, in turn, affect the stem cells. Both are difficult to control, but can provide a more realistic view.

Based on these paradigms, it is worth briefly considering how the non-definitive stem cells systems (Fig. 1), ES and iPS cells, fit into predictive stem cell toxicity testing. At the present time, these cells are used to produce functionally, mature lineage-specific cells such as hepatocytes, cardiomyocytes and neurons. These and other cell types can be produced in larger numbers and presumably at a lower cost than their primary counterparts. Embryonic stem cells are used as an *in vitro* developmental toxicity model to predict teratogenicity. The use of ES and/or iPS cells for definitive stem cell system toxicity testing is certainly on the horizon. It should be remembered however, that even to produce functionally, mature hepatocytes, cardiomycytes and other cells, the ES and iPS cells must pass through the definitive stem cell compartment specific for the cells being produced. In other words, the ES and iPS cells should produce an organization analogous to that shown in Fig. 2. If this transpires, then the face of toxicity testing, and stem cell toxicity testing in particular, as well as many other applications, could significantly change the face of biological and toxicological research in the future.

Acknowledgements

The authors would like to thank Drs. Patricia Wood and William Hrushesky at the Dorn Research Institute of the William Jennings Bryan Dorn Veterans Affairs Medical Center in Columbia, South Carolina for sharing their insights and knowledge of circadian rhythms that led to the results shown in Section 5F.

Author details

Holli Harper and Ivan N. Rich

HemoGenix, Inc, U.S.A.

References

[1] Horwitz EM, Le Blanc K, Dominici M, Mueller I, Slaper-Cortenbach I, Marini FC, Deans RJ, Krause DS, Keating A; The International Society for Cellular Therapy:

Clarification of the Nomenclature for MSC: the International Society for Cellular Therapy Position Statement. Cytotherapy 2005; 7:393-395.

[2] Potten, CS. Cell cycles in cell hierarchies. Int J Radiat Biol Relat Stud Phys Chem Med. 1986; 49:257-278.

[3] Lemischka IR. The haematopoietic stem cell and its clonal progeny: mechanism regulating the hierarchy of primitive haematopoietic cells. Cancer Surv. 1992; 15:3-18.

[4] Yahata, T, Muguruma, Y, Yumino, S, Sheng, Y, Uno, T, Matsuzawa, H, Ito, M, Kato, S, Hotta, T, Ando, K. Quiescent human hematopoietic stem cells in the bone marrow niches organize the hierarchical structure of hematopoiesis. Stem Cells. 2008; 26: 3228-3236.

[5] Campbell, CJ, Lee, JB, Levadoux-Martin, M, Wynder, T, Xenocostas, A, Leber, B, Bhatia, M. The human stem cell hierarchy is defined by a functional dependence on Mcl-1 for self-renewal capacity. Blood, 2010; 116:1433-1442.

[6] Levesque, JP, Winkler, IG. Hierarchy of immature hematopoietic cells related to blood flow and niche. Curr. Opin. Hematol. 2011; 18: 220-225.

[7] Staal FJ, Baum C, Cowan C, Dzierzak E, Hacein-Bey-Abina S, Karlsson S, Lapidot T, Lemischka I, Mendez-Ferrer S, Mikkers H, Moore K, Moreno E, Mummery CL, Robin C, Suda T, Van Pel M, Vanden Brink G, Zwaginga JJ, Fibbe WE. Stem cell self-renewal; lessons from bone marrow, gut and iPS towards clinical applications. Leukemia. 2011; 25:1095-1102.

[8] Van Zant G, de Haan G, Rich IN. Alternatives to stem cell renewal from a developmental viewpoint. Exp Hematol. 1997; 25:187-192.

[9] Kola I, Landis J. Can the pharmaceutical industry reduce attrition rats? Nature Reviews: Drug Discovery. 2004; 3:711-715.

[10] Mahajan R, Gupta K. Food and drug administration's critical path iniative and innovations in drug development paradigm: Challenges, progress, and controversies. J Pharm Bioallied Sci 2010; 2: 307-313.

[11] National Cancer Institute, Division of Cancer Treatment and Diagnosis, Developmental Therapeutics Program. http://dctd.cancer.gov/ProgramPages/dtp/tools_drug_discovery.htm.

[12] Yamori T. Panel of human cancer cell lines provides valuable database for drug discovery and bioinformatics. Cancer Chemother Pharmacol. 2003. 52 Suppl 1:S74-S79.

[13] Shanks N, Greek R, Greek L. Are animal models predictive for humans? Philos Ethics Humanit Med. 2009; 15: 2.

[14] Toxicity Testing in the 21st Century. A Vision and a Strategy. Committee on toxicity testing and assessment of environmental agents. National Research Council of the

National Academies. National Academic Press. ISBN 0-309-10993-0. 2007. http://www.nap.edu/catalog/11970.html.

[15] Senior JR. Monitoring for hepatotoxicity: what is the predictive value of liver "function" tests? Clin Pharmacol Ther 2009; 85: 331-334.

[16] Maziasz T, Kadmabi VJ, Silverman L, Fedyk E, Alden CL. Predictive toxicology approaches for small molecule oncology drugs. Toxicol Pathol 2012; 38: 148-164.

[17] Firestone M, Kavlock R, Zenick H, Kramer M. The U.S. environmental protection agency strategic plan for evaluating the toxicity of chemicals. J Toxicol Environ Health B Crit Rev. 2010; 13: 139-162.

[18] Benigni R, Bossa C. Mechanism of chemical carcinogenicity and mutagenicity: a review with implications for predictive toxicology. Chem Rev. 2011; 111: 2507-2536.

[19] Gleeson MP, Modi S, Bender A, Robinson RL, Kirchmair J, Promkatkaew M, Hannonqbua S, Glen RC. The challenges involved in modeling toxicity data in silico: a review. Curr Pharm Des. 2012; 18: 1266-1291.

[20] REACH Directive, http://ec.europa.eu/environment/chemicals/reach/reach_intro.htm

[21] Bradley TR, Metcalf D. The growth of mouse bone marrow cells in vitro. Aust J Exp Biol Med Sci. 1966; 44: 287-299.

[22] Pluznik DH, Sachs L. The induction of clones of normal mast cells by a substance from conditioned medium. Exp Cell Res. 1966; 43: 553-563.

[23] Cole RJ, Paul J. The effects of erythropoietin on haem synthesis in mouse yolk sac and cultured foetal liver cells. J Embryol Exp Morphol. 1966; 15: 245-260.

[24] Stephenson JR, Axelrad AA, McLeod DL, Shreeve MM. Induction of colonies of haemoglobin-synthesizing cells by erythropoietin in vitro. Pro Natl Acad Sci USA. 1971; 68:1542-1546.

[25] Iscove NN, Sieber F, Winterhalter. Erythroid colony formation in cultures of mouse and human bone marrow: analysis of the requirement for erythropoietin by gel filtration and affinity chromatography on agarose-concanavalin A. J Cell Physiol. 1974; 83:309-320.

[26] Pessina A, Albella B, Bueren J, Brantom P, Casati S, Gribaldo L, Croera C, Gagliardi G, Foti P, Parchment R, Parent-Massin D, Sibiril Y, van Den Heuvel R. Prevalidation of a model for predicting acute neutropenia by colony forming unit granulocyte/macrophage (CFU-GM) assay. Toxicol In Vitro. 2001; 15: 729-740.

[27] Pessina A. Albella B, Bayo M, Bueren J, Brantom P, Casati S, Croera C, Gagliardi G, Foti P, Parchment R, Parent-Massin D, Schoeters G, Sibiril Y, Van Den Heuvel R, Gribaldo L. Application of the CFU-GM assay to predict acute drug-induced neutropenia: an international blind trial to validate a prediction model for the maximum tolerated dose (MTD) of myelosuppressive xenobiotics. Tox Sci. 2003; 75:355-367.

[28] Pessina A, Parent-Massin D, Albella B, Van Den Heuvel R, Casati S, Croera C, Malerba I, Sibiril Y, Gomez, S, de Smedt A, Gribaldo L. Application of human CFU-Mk assay to predict potential thrombocytotoxicity of drugs. Toxicol In Vitro. 2009; 23:194-200.

[29] Bacigalupo A, Piaggio G, Podesta M, Figari O, Benvenuto F, Sogno G, Tedone E, Raffp MR, Grassia L, Ferrero R, et al. Bone Marrow Transplant. 1995; 15:221-226.

[30] Jansen EM, Hanks, SG, Terry C, Akard LP, Thompson JM, Dugan MJ, Jansen J. Prediction of engraftment after autologous peripheral blood progenitor cell transplantation: CD34, colony-forming unit granulocyte-macrophage, or both? Transfusion. 2007; 47:817-823.

[31] Gluckman E. Milestones in umbilical cord blood transplantation. Blood. 2011; 25:255-259.

[32] Rich IN, Hall KM. Validation and development of a predictive paradigm for hemotoxicology using a multifunctional bioluminescence colony-forming proliferation assay. Toxicol Sci. 2005; 87: 427-441.

[33] Spellman S, Hurley CK, Brady C, Phillips-Johnson L, Chow R, Laughlin M, McMannis J, Reems JA, Regan D, Rubinstein P, Kurtzberg J. Guidelines for the development and validation of new potency assays for the evaluation of umbilical cord blood. Cytotherapy. 2011; 13: 848-855.

[34] Rich IN, Kubanek B. The effect of reduced oxygen tension on colony formation of erythropoietic cells in vitro. Br. J Haematol. 1982; 52: 579-588.

[35] Rich IN. The role of the macrophage in normal hemopoiesis. II. Effect of varying physiological oxygen tensions on the release of hemopoietic growth factors from bone marrow-derived macrophages in vitro. Exp Hematol. 1986; 14: 746-751.

[36] Hall KM, Rich IN. Bioluminescence assays for assessing potency of cellular therapeutic products. In: Cellular Therapy: Principles, Methods and Regulations. ISBN: 978-1-56395-296-8. Bethesda, MD. 2009. p 581-591.

[37] Hall KM, Harper H, Rich IN. Hematopoietic Stem Cell Potency for Cellular Therapeutic Transplantation. In: Advances in Hematopoietic Stem Cell Research, Rosana Pelayo (Ed.), ISBN: 978-953-307-930-1, InTech, 2012. http://www.intechopen.com/books/advances-in-hematopoietic-stem-cell-research/hematopoietic-stem-cell-potency-for-cellular-therapeutic-transplantation.

[38] Rich IN. In vitro hematotoxicity testing in drug development: a review of past, present and future applications. Curr Opin Drug Discov Devel. 2003; 6:100-109.

[39] Rich IN. High-throughput in vitro hemotoxicity testing and in vitro cross-platform comparative toxicity. Expert Opin Drug Metab Toxicol. 2007; 3: 295-307.

[40] FDA Guidance for Industry, Bioanalytical Method Validation, 2001. http://www.fda.gov/downloads/Drugs/.../Guidances/ucm070107.pdf.

[41] Guidance Document on Using In Vitro Data to Estimate In Vivo Starting Doses for Acute Toxicity. 2001. NIH Publication No.: 0.1-4500. http://iccvam.niehs.nih.gov/docs/acutetox_docs/guidance0801/iv_guide.pdf.

[42] FDA Guidance for Industry and Reviewers. Estimating the Safe Starting Dose in Clinical Trials for Therapeutics in Adult Healthy Volunteers. 2002. http://www.fda.gov/downloads/Drugs/.../Guidances/UCM078932.pdf.

[43] Mann HJ. Drug-associated disease: cytochrome P450 interactions. Crit Care Clin 2006; 22:329-345.

[44] The Cytochrome P-450 Enzyme System. http://www.edhayes.com/startp450.html.

[45] Monte-Silva K, Fuo M-F, Thirugnanasambandam N, Liebetanz D, Paulus W, Nitsche MA. Dose-dependent inverted U-shaped effect of dopamine (D2-like) receptor activation on focal and nonfocal plasticity in humans. J Neurosci. 2009; 29: 6124-6131.

[46] Celik I, Sürücü O, Dietz C, Haymach JV, Force J, Höschele I, Becker CM, Folkman J, Kisker O. Therapeutic efficacy of endostatin exhibits a biphasic dose-response curve. Cancer Res 2005; 65: 11044-11050.

[47] Innominato PF, Levi FA, Bjarnason GA. Chronotherapy and the molecular clock: Clinical implications in oncology. Adv Drug Deliv Rev 2010; 61: 979-1001.

[48] Takeda N, Maemura K. Circadian clock and cardiovascular disease. J Cardiol. 2011; 57: 249-256.

[49] Cutolo M. Chronobiology and the treatment of rheumatoid arthritis. Curr Opin Rheumatol. 2012; 24: 312-318.

[50] Haus E, Lakatua DJ, Swoyer J, Sackett-Lundeen L. Chronobiology in hematology and immunology. Am J Anat. 1983; 168: 467-517.

[51] Laerum OD. Hematopoiesis occurs in rhythms. Exp Hematol. 1995; 23: 1145-1147.

[52] Mendez-Ferrer S, Chow A, Merad M, Frenette PS. Circadian rhythms influence hematopoietic stem cells. Curr Opin Hematol. 2009; 16: 235-242.

[53] Naitoh P, Englund CE, Ryman DH. Circadian rhythms determined by cosine curve fitting: Analysis of continuous work and sleep-loss data. Behav Res Metho Instrum Comp. 1985; 17: 630-641.

[54] Dogliotti L, Tampellini M, Levi F. Chronochemotherapy of colorectal cancer. Biological Clock: Mechanisms and Applications, Touitou Y (ed); 1998; 475-481.

[55] Liebsch M, Grune B, Seller A, Butzke D, Oelgeschlaeger M, Pirow R, Adler S, Riebeling C, Luch A. Alternatives to animal testing: current status and future perspectives. Arch Toxicol. 2011; 85: 841-858.

[56] Olaharski AJ, Uppal H, Cooper M. Platz S, Zabka TS, Kolaja KL. In vitro to in vivo concordance of a high throughput assay of bone marrow toxicity across a diverse set of drug candidates. Toxicol Let. 2009; 188: 98-103.

[57] Zang JH, Chung TDY and Oldenburg Kr: A simple statistical parameter for use in evaluation and validation of high throughput screening assays. J Biomol Screen 1999; 4: 67-73

Section 2

Stem Cells in Disease

Chapter 7

Stem Cell-Based (Auto)Grafting: From Innovative Research Toward Clinical Use in Regenerative Medicine

Bela Balint, Slobodan Obradovic, Milena Todorovic,
Mirjana Pavlovic and Biljana Mihaljevic

Additional information is available at the end of the chapter

http://dx.doi.org/10.5772/54473

1. Introduction

In brief, adult stem cells (SCs) give rise to repopulation (engraftment) of recipient's bone marrow (BM) followed by complete and long-term reconstitution of hematopoiesis. In addition, totipotent SCs are also capable of colonizing different tissues (homing). Initial studies showed that "implantation" of autologous SCs into damaged and ischemic area induces their homing and subsequent "transdifferentiation" into the cell lineages of host organ, including collateral vessel formation. Angiogenesis growth factors – or genes encoding these proteins – promote the development of collateral micro-angiogenesis or "therapeutic neovascularization" [1– 5].

Generally, SC transplant involves the administration of high-dose chemotherapy (conditioning regimen) and (re)infusion of collected cells in order to obtain an abolition of disease, as well as to get hematopoietic reconstitution and clinical improvement of patient. SC transplant with reduced-intensity conditioning (RIC) can be offered to patients who are ineligible for high-dose conditioning because of their age or comorbidities [2]. Hematological diseases have so far been the most common indication of this treatment modality; it has been less often used for nonmalignant disorders. Nowadays BM and peripheral blood (PB) derived SC transplants are more common in adult allogeneic or autologous setting [2, 6– 8]. Umbilical cord blood (UCB) transplants have provided hopeful results in pediatric setting mainly when a matched unrelated SC donor is not obtainable [9–12].

In clinical practice, SCs can be collected by: (a) multiple aspirations from BM; (b) harvesting PB after mobilization with chemotherapy and/or growth factors (rHuG-CSF), and (c) by specific processing from UCB. SCs collected from the stated sources can be clinically applied (transplanted) immediately following harvesting (allogeneic setting) or after a long-term storage in frozen state – cryopreservation (autologous setting) [2].

2. Stem cell transplants – A short chronological consideration

Independent SC-researchers recognized that all blood cells originate from one primitive BM cells (totipotent SC) located in marrow – space where the entire hematopoiesis takes place. Initial animal studies revealed that the BM was the organ most sensitive to the damaging effect of gamma irradiation [2]. Quickly, it became clear that (re)infusion of marrow cells or SCs could rescue lethally irradiated animals. Thomas with colleagues started on, and after that optimized (for this initial period) the BM transplant (BMT) program for humans and published in 1957 the first clinical results [13]. During the late 1950s have been also described the first syngeneic BMT in patients with leukemia [14]. Mathe and coworkers published the treatment of patients by allogeneic SC transplant after of accidental irradiation [15]. These transplants were performed before the discovery of major histocompatibility (MCH) system. In addition, it is not excluded that the observed recovery of some patients were a result of the recovery of autologous hematopoietic system [16]. The first successful BMT (allogeneic) was performed on a child with severe combined immunodeficiency disease (SCID). Cells were collected from his sister, and his immune system was restored following transplant [2]. However, in most cases transplants in humans have been unsuccessful (because of the graft rejection or expansion of the Graft versus Host Disease – GvHD).

The modern era of SC transplants – as a standard therapy – started with fundamental discovery and permanent progress in the knowledge of MHC, that is human leukocyte antigen (HLA) system [17]. These antigens give the body's immune system the ability to determine what belongs and what does not belong to the human body. Whenever the immune system does not recognize antigens expressed on a cell surface, it produces antibodies and other mediators to destroy the cells with non-recognizable antigens. In order for BMT to work, the recipient's immune system must not try to destroy the donated cells. This comprises that the HLA antigens on the donated SCs have to be identical or extremely similar to the antigens of the recipient's cells. Even with this careful HLA matching, transplant may still fail because recipient's immune system destroys transplanted cells (graft rejection) or donor's cells attempt to damage recipient's target cells (GvHD) [2]. Thomas and coworkers almost immediately published positive results of the first allogeneic BMT in patients with hematologic malignancies – using cells from donors selected on the basis of the HLA system [18]. After all, Thomas ED was awarded the Nobel prize in medicine (1990) for his overall pioneering work on BMT topic. He was awarded the prize because of his numerous triumphant activities in both, experimental and clinical transplant setting.

In addition to marrow, PB has gained popularity as a SC source since their initial introduction in the early 1980s [19]. Over the past decades, the use of these transplants has expanded rapidly [6– 8]. Using umbilical cord blood (UCB) derived SCs, successful transplant occurred the first time in the treatment of Fanconi anemia and other disorders later than [9–12]. It is known that

http://dx.doi.org/10.5772/54473

only about one-third of patients have related HLA-matched donor. For that reason, some sources of allogeneic donors – including unrelated HLA-compatible individuals, have to be considered as the possible alternative. As a result, National Marrow Donor Program's registries of volunteer donors has been created and data accumulated by organizing a unique database for potential donors (Bone Marrow Donors Worldwide – BMDW) [20].

The late 1990s brought a new apprehension regarding the biology and related novel clinical potential of SCs. Researchers began to realize that manipulation of adult animal tissues could sometimes yield previously unsuspected cell types; for example, that some BM derived SCs could be turned into cardiomyocytes, hepatocytes or nerve and other somatic cells – phenomenon known as the SC "plasticity". Finally, using SC plasticity, cell–based therapies for treatment of the ischemic heart diseases started through beginning of the new millennium and currently are in an expansion phase in the other fields of regenerative medicine [21–30].

3. Adult stem cells: New concepts in phenotypes and functionality

To prove that SCs derived from BM and PB, including hematopoietic SCs, are indeed trans-differented or transformed into solid organ specific cells, several conditions must be met:

- The origin of the exogenous cell integrated into solid-organ time must be documented by marking the cell, preferably at the single-cell level;
- Cell should be processed with a minimum of *ex vivo* manipulation which may make them more susceptible to crossing lineages;
- The exogenous cells must be shown to have become an integral morphological part of the newly acquired tissue;
- Transdifferented cells must have shown to acquire the function of the particular organ into which it has been integrated both, by expressing organ-specific proteins and by showing specific organ function.

Nevertheless, taking into consideration their common features described in the literature, it is very likely that various investigators have described overlapping populations of developmentally early SCs that are closely related. Our intention is to make a clear distinction between three different types of adult SCs.

4. The concept of hematopoietic stem cells

Organ/tissue specific niche (like in BM, liver, etc) exists as a deposit (storage) of the adult SCs in a specific location [31]. These cells are circulating in a very low number in the PB. Accumulating evidence suggests that SCs may also actively migrate/circulate in the postnatal period of life. SC trafficking/circulation may be one of the crucial mechanisms that maintains the pool of SCs dispersed in SC-niches of the same tissue, that are spread throughout different anatomical areas of the body. This phenomenon is very well described for hematopoietic SCs (HSCs), but other, already tissue committed SC or TCSC (for example, endothelial, skeletal muscle or neural SCs) are probably circulating as well. BM is the home of migrating SCs with not only HSCs within their niches, but also a small number of TCSC, which might be the reason why many authors think that HSC may transdifferentiate, although we do not have a direct proof for that. They might have plasticity, but not necessarily the "potential for transdifferention" [32–39]. What is differentiated in the tissue of injection might be TCSC characteristic for that tissue. It has been shown that number of these cells is decreased with ageing (long living and short living mice and humans). It would be interesting to identify genes that are responsible for tissue distribution/expansion of TCSC. These genes could be involved in controlling the life length of the mammals.

Therefore, BM derived SCs are a heterogeneous population of cells with HSC and TCSC, the morphological and functional characteristics of which are different from HSC. Their number among mononuclear cells (MNCs) is very low (approximately one cell per 1 000 – 10 000 marrow MNCs) within young mammals and might play a role in healing of small injuries [31, 32].

In severe injuries (like hart infarct or stroke) they have no possibility to reveal their full therapeutic potential. The allocation of these cells to the damaged areas depends on homing signals that maybe inefficient in the presence of some other cytokines or proteolytic enzymes that are released from damaged tissue associated leukocytes and macrophages. We can envision, for example that metalloproteinases released from inflammatory cells may degrade SDF-1 locally, and thus perturb homing of $CXCR4^+$ TCSC. There is possibility that these cells while "trapped" in BM are still in: "latent stage" – not fully functional and need the appropriate activation signals by up till now unknown factors [32–37].

These cells also, at least in some cases could be attracted to the inflammatory areas, and if not properly incorporated into the damaged tissue they may transform and initiate tumor growth. Briefly, between the pools of TCSCs, there are probably those already committed to transdifferentiate into neural cells, or cells of tissues and organs other ten neural, but we still do not have the control over their tracking, homing and finally regenerative capacity in the given tissue, which is a fundamental prerequisite for successful regenerative therapy.

5. The concept of "very small embryonic-like" stem cells

In a discovery that has the potential to change the face of SC research, a University of Louisville scientist has identified cells in the adult body that seem to behave like embryonic SCs [38, 40–44]. The cells, drawn from adult BM, look like embryonic SCs and appear to mimic their ability to multiply and develop into other kinds of cells. The finding, presented the first time at the 47th Annual Meeting of the American Society of Hematology (ASH), was announced at the society's news conference. A study by Ratajczak's team published in the journal "Leukemia" was the first to identify a type of SC in adult marrow that acts differently than other BM derived

SCs [38]. The newly identified cells – called "Very Small Embryonic-Like" (VSEL) SCs – have basically the same ultrastructure and protein markers as embryonic SCs [38]. Ratajczak and several other researchers from mentioned ASH meeting showed that VSEL SCs mobilize into the blood stream to help repair damaged tissue following a stroke [37]. In further research advance, Ratajczak's team also has grown VSEL cells in a lab and has stimulated them to change into nerve, heart and pancreas cells [40–42]. The differences in ultrastructure between HSC and VSELs are shown in Figure 1.

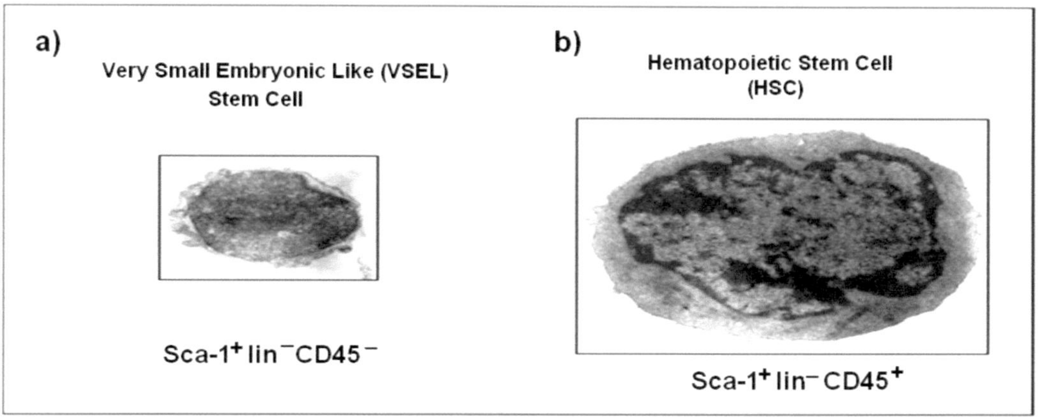

(For picture thanks to kind permission of M. Ratajczak)

Figure 1. Ultra-structural differences between mouse VSEL and HSC

Along with this new concept, there is a premise that in regenerative therapy done before, with HSCs (considered to have plasticity and multipotency) the VSELs were "contaminants" that actually contributed to positive regenerative clinical outcome, since they have those capabilities. This is an interesting concept which should be seriously considered in humans. However, since VSELs have been found in human UCB and BM first [37], and then used and applied with the patients [43–44] they seem to be of a critical importance for consideration of SC transplant choice based upon the phenotype and number of SCs aimed to be transplanted within a given clinical scenario.

6. The concept of mesenchymal stem cell – with dental pulp cells as an example

Many human tissues are the source of SCs responsible for tissue development and regeneration. Beside marrow (Bone Marrow Stromal Stem Cells – BMSCs), currently it is considered that dental pulp is practically the most approachable and the most important source of adult mesenchymal SCs [45–47] (Figure 2).

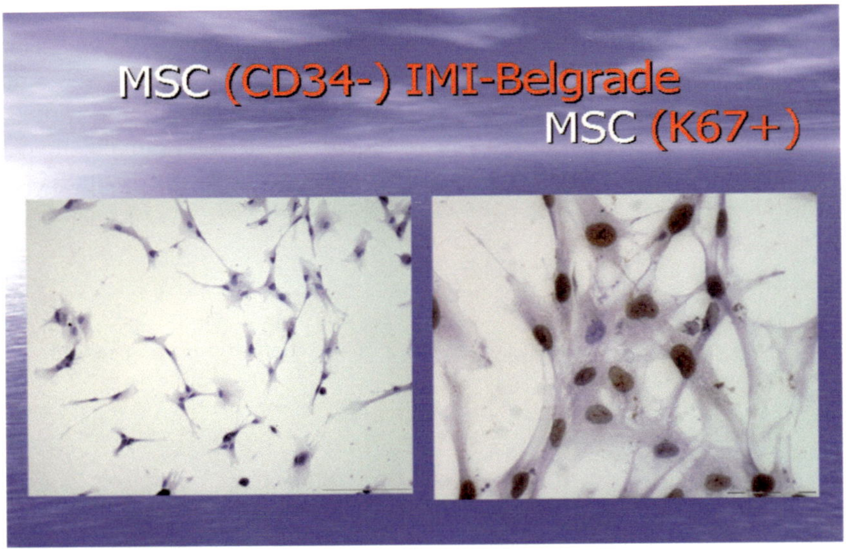

(With courtesy of V. Todorovic)

Figure 2. Mesenchymal stem cells

Within the last decade, several populations of SCs from dental pulp were isolated and characterized: a) Dental Pulp Stem Cells – DPSC; b) cells from Human Exfoliated Decidual teeth – SHED; and c) Immature Dental Pulp Cells – IDPC [46, 47]. These cells are of the ectomesenchymal origin, located in perivascular niche, highly proliferative, clonogenic, multipotent and similar to BMSCs. Within *in vitro* conditions, they can differentiate with certain intercellular differences toward odontoblasts, hondrocytes, osteoblasts, adipocytes, neuron/glial cells, smooth and skeletal muscle cells. Within *in vivo* conditions (after implantation) they show different potential for dentine formation, as well as osteogenesis; after transplant in mouse with compromised immune system, they make good grafts in different tissues, and are capable of migrating into the brain, where they survive a certain time while reaching neurogenic phenotype. DPSCs have immunomodulatory effect, as they can be involved into immune response during infection of dental pulp by NF-kB activation and by inhibiting T-cell proliferation, suggesting their immunosuppressive effect [47].

The future research should give us the complex data on the molecular and functional characteristics of dental pulp SCs, as well as differences between various populations of these cells. Such research would fundamentally contribute to the better knowledge on the dental pulp SCs, which is necessary due to their potential clinical application in *in vivo* cell transplant, tissue engineering, and gene therapy (*in vivo* or *ex vivo*). Actually, by the isolation of IDPCs, which are the most primitive, but also the most plastic (similar to embryonic SCs) they are opening the new perspectives in a potential therapeutic application of these cells not only in regeneration of dentine, otherwise also the regeneration of periodontal and "bone-junction" tissue of craniofacial region, as well as in the therapy of neurotrauma, myocardial infarction and other tissue damages [46, 47].

7. Stem cell harvesting and *ex vivo* manipulation

For therapeutic use, SCs can be collected by: a) multiple aspirations from BM; b) harvesting from PB after mobilization (chemotherapy and/or growth factors – rHuG-CSF); and c) collection from UCB. Collected cells can be clinically applied (transplanted) immediately following harvesting (allogeneic setting) or after storage in frozen state or cryopreservation (autologous setting).

8. Bone marrow derived stem cells

Historically, BM was the first SC source for transplants. Cells were collected by multiple aspirations from the iliac crests, under sterile conditions, while the donor was generally anesthetized. The target volume of collected BM aspirate is 10 – 15 mL per kg of donor body mass (kgbm). In order to provide required number of total nucleated cells (TNCs) – that is TNC ≥ $3x10^8$/kgbm – around 200 aspirations are required (single aspirate volume = 2 – 5 mL) (Figure 3).

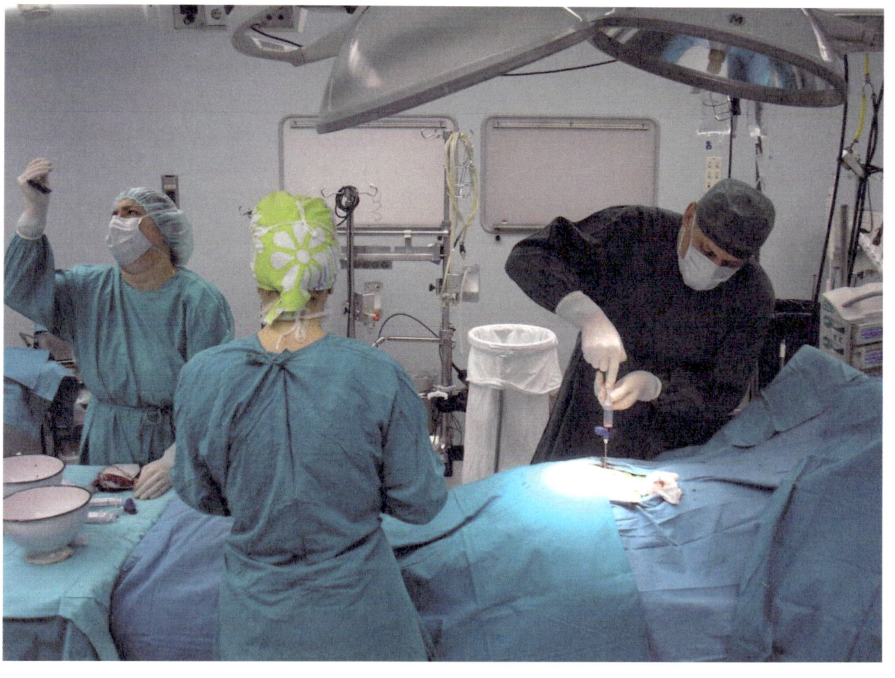

Figure 3. Stem cell collection from bone marrow

After collection, BM aspirate should be filtered in order to remove bone and lipid particles and cell aggregates. Anticoagulation is created using citrate solution and by using of the heparin diluted in saline (5 000 IU/500 mL) [2, 48–50].

BM aspirate volume – precisely red blood cell count or plasma quantity – reduction is required (processing), especially for ABO incompatible transplants. Depletion of T-cells in cell suspension is achieved using *ex vivo* purging (by immunomagnetic technique). These SC purification procedures (processing and purging) enable reduction of red cell for around 80 – 90% and depletion of T-cells with 3 – 4 Log_{10} [50].

9. Peripheral blood derived stem cells

The PB derived SC transplants are characterized by: a) less invasive cell collection; b) lack of the risks of general anesthesia; c) rapid hematopoietic reconstruction; d) low harvest volume (200 – 300 mL), and e) inferior transplant-related morbidity. Thus, the number of patients treated by PB derived SCs is ever increasing, especially in autologous transplant setting [6–8, 48–51].

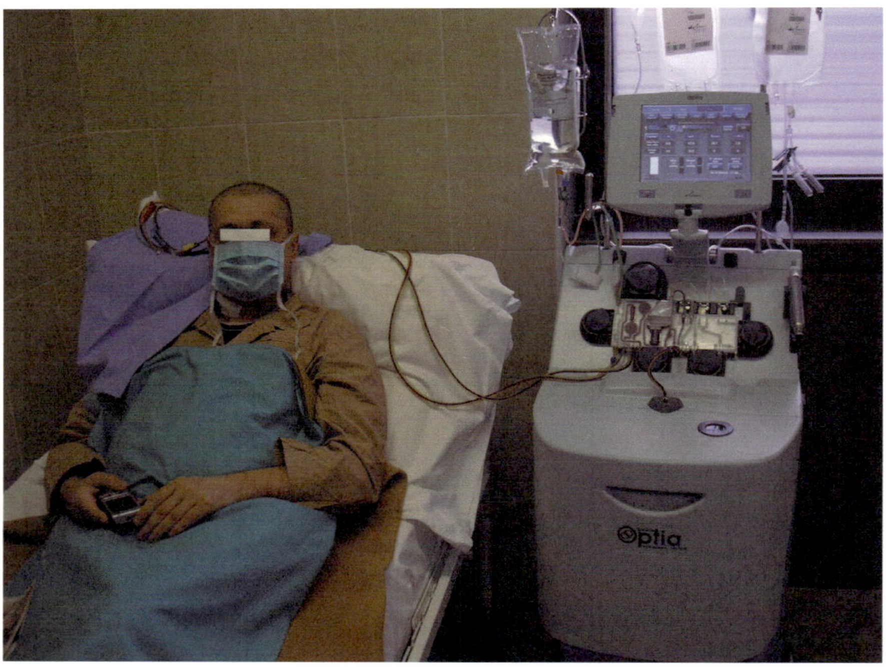

Figure 4. Stem cell harvesting from peripheral blood

In steady state hematopoiesis SCs, that is $CD34^+$ cells are in very low proportion (0.01% to 0.1% compared to MNCs) in PB, but they can be mobilized from BM. Allogeneic donors are given rHuG-CSF 5 (10 μg/kgbm per day). The count of $CD34^+$ cells in the circulation begins to rise after 3rd day, and peaks on the 5th day of rHuG-CSF administration. In autologous setting, patients are given higher rHuG-CSF dose (12-16 μg/kgbm or more daily) combined with chemotherapy [8, 48–51].

In allogeneic setting, the first SC collection is performed typically on the 5th day (Figure 4.). The optimized timing of an autologous SC harvesting is more complex and controversial. The leukocyte count commonly does not correlate with the circulating $CD34^+$ number. The count of circulating $CD34^+$ cells evidently correlate with the superior $CD34^+$ yield in harvest. Generally, at PB $CD34^+$ count ≅ 10/μL, expected SC yield ≅ $1x10^6$ per kgbm. It is also presented that for a $CD34^+ \geq$ 20-40/μL of PB the possibility of the $CD34^+ \geq 2.5x10^6$ per kgbm is 15% using one standard apheresis, and 60% after one large-volume SC harvesting [6–8, 49].

The target CD34+ cells should be $330x10^6$ per unit or ≥ $2\text{-}4x10^6$/kgbm of the recipient in order to expect successful transplant. Recent data support a benefit associated with greater $CD34^+$ yield (≥ $5x10^6$/kgbm) compared to the minimum required cell quantity for engraftment (≥ 1 x 10^6/kgbm) in autologous setting [20]. Finally, results obtained in our SC transplant center confirmed that large-volume apheresis is efficient ($CD34^+ \geq 5x10^6$/kgbm) if the circulating $CD34^+$ count was around 40-60/μL after mobilizing regiment [8].

10. Umbilical cord blood derived stem cells

Patient's request for SCs have only in ≤ 30% (related) and ≤ 85% (unrelated) possibility of finding an adult allogeneic donor [20]. Because of the limited availability of donors, attention has turned to alternative sources of HLA-typed SCs. In recent years, UCB has emerged as a feasible alternative source of transplantable $CD34^+$ cells for allogeneic transplant, mainly in patients who lack HLA-matched donors of BM or PB derived SCs [9–1 2].

SCs obtained from the UCB immediately after birth are usually referred to as neonatal SCs. These cells are less mature than those in BM. The advantage of the use of UCB is painless and non–invasive collection. UCB has advantage that – despite its high content of immune cells – it does not produce severe GvHD. Precisely, the "naive" nature of UCB lymphocytes permits the use of partly HLA-mismatched grafts without higher risk for severe GvHD relative to BM transplant from a full matched unrelated donor. Thus, UCB grafts do not need to be as "rigorously" matched to a recipient as BM or PB grafts [10–12].

On the contrary, the major disadvantage of this cell source is the limited number of SCs. UCB volume is typically 80 – 200 mL, with a TNC count ≈$1x10^9$ and approximately $CD34^+$ count ≈$3x10^6$ per unit. Thus, UCB is an accepted cell source for pediatric patients and for whom a matched unrelated BM or PB cell donor is unavailable. However, a higher risk of graft failure was noticed in children weighing ≥ 45 kg. Since the number of SCs in UCB is limited and the collection can occur only in a single occasion – its use in adult patients can be more problematic. Finally, since SCs in the UCB are "more primitive", the engraftment process takes longer with UCB, leaving the patient vulnerable to posttransplant infections or bleeding. However, "more primitive" SCs in UCB have the potential to give rise to non-hematopoietic cells (myocardial, neural and endothelial cells, etc) by transdifferentiation [2, 9–12].

11. Stem cell cryopreservation practice

Efficient transplant program requires both, high-quality harvesting and cryopreservation techniques for obtaining adequate yield and recovery of the SCs. In practice, SC cryopreservation consists of the following steps: a) aspirate processing; b) equilibration (cell exposure to cryoprotective agent) and freezing; c) storage at temperature –80±5°C (mechanical freezer), at –140±5°C (mechanical freezer or steam of nitrogen) or at -196°C (liquid nitrogen); and d) cell thawing in a water bath at 37±3°C. There are several cryopreservation protocols, using primarily DMSO in autologous plasma. The optimal cooling velocity in controlled-rate cryopreservation setting is –1°C. The transition from liquid to solid phase is also critical period – because of released fusion heat – since a significant reduction in cell recovery and viability has been observed when this period is prolonged. The compensation of the released fusion heat is required, using elevated cooling rate (–2 °C per minute) during transition period. Finally, there is data showing that uncontrolled-rate freezing is also useful in SC cryopreservation [48–52].

Cryopreserved SCs are thawed rapidly in a water bath at 38±2°C at the patient's bedside and infused immediately through a central vein catheter. Generally, patients tolerate the infusion of unprocessed SCs well, with no side effects. The grade of the potential reinfusion–related toxicity is associated with total DMSO quantity in the thawed cell concentrate. Alternatively, cryoprotectant can be removed by washing after thawing, but this procedure results in substantial cell loss.

12. Conventional stem cell transplants – A synopsis of the clinical practice

Generally, SC transplants include the use of high-dose chemotherapy in order to obtain disease eradication and (re)infusion (allogeneic transplant or autologous SC support) of cells collected to get hematopoietic and clinical renewal. SC transplant with reduced-intensity conditioning (RIC) can be offered to patients who are disqualified for high-dose conditioning because of their age or comorbidities. Malignant hematological diseases and some immune-mediated disorders (Table 1) are the most common indication of this therapeutic approach using SCs [2, 50].

The efficacy of transplants depends on the type of disease, its stage and sensitivity for chemotherapy, patients' age and general condition, as well as degree of the HLA-matching. In general, survival rates are around 30 – 60% for otherwise fatal diseases. Details of the SC clinical use – that is optimized treatment timing and efficacy, peritransplant complications, etc. – of the transplants in presented hematological disorders will not discussed in this paper.

Briefly, autoimmune diseases, which do not respond to standard immunosuppression, could benefit from immunoablative therapy. The idea of treatment of immune-mediated disorders (e.g. multiple sclerosis) by autologous SC transplant is based on the hypothesis that immunoablative treatment can destroy the patients "anti-self-lymphocytes" (i.e. an "immune-resetting" process). The beneficial immunomodulating effect of allogeneic SC transplant in

http://dx.doi.org/10.5772/54473

therapy of hematological malignancies has long been known, but only recently have systems been developed to separate the GvL effect from GvHD. Using donor-specific lymphocytes, the best results were obtained in chronic myelogenous leukemia. Moderate successes have been reported in relapsing acute myeloid leukemia, myelodysplastic syndrome, multiple myeloma and some responses were obtained for acute lymphoid leukemia [2, 8, 50].

BM malignant or dysplastic disorders
Leukemias
Hodgkin's and non-Hodgkin's lymphoma
Multiple myeloma
Myelodysplastic/myeloproliferative disorders
Benign immune mediated disorders
Severe combined immunodeficiency disease
Marrow failure syndromes
Severe aplastic anemia
Autoimmune disorders
Thalassemia
Congenital Immune deficiencies
Solid tumors
Breast cancer
Ovarian cancer
Testicular cancer
Wilm's tumor
Neuroblastoma
Rhabdomyosarcoma
Ewing sarcoma

Table 1. Current indications and relative suggestions for SC transplant

Although SC transplant-related mortality and morbidity have reduced, SC transplants continue to pose numerous potential complications. The most frequent complications are even now engrafting failure, virus or opportunistic infections and acute or chronic GvHD. Less toxic transplants, in the form of non-myeloablative conditioning regimens, are being actively investigated, with the promise of expanding indications for allogeneic transplant. In addition, SC transplant with RIC can be offered to patients who are disqualified for high-dose conditioning because of their age or comorbidities. A careful proactive assessment to identify, treat, and, hopefully, prevent adverse events is essential to a successful transplant [2].

We have previously analyzed our results of PB vs. BM derived SC transplants based on the hematopoietic reconstitution. Transplants were used for the treatment of patients with severe aplastic anemia, acute lymphoblastic leukemia, acute non–lymphoblastic leukemia, chronic myeloid leukemia, multiple myeloma, Hodgkin's and non–Hodgkin's lymphoma, breast and ovarian cancer, extragonadal non–seminal germ cell tumor, and severe multiple sclerosis. The $CD34^+$ yields for allogeneic and autologous transplants were eminent: 16.7±9.8 × 10^6/kgbm and 11.8±6.1 × 10^6/kgbm, respectively [8, 26, 27]. A typical histogram with high–level of the $CD34^+$ ratio in obtained PB harvest is presented in Figure 5.

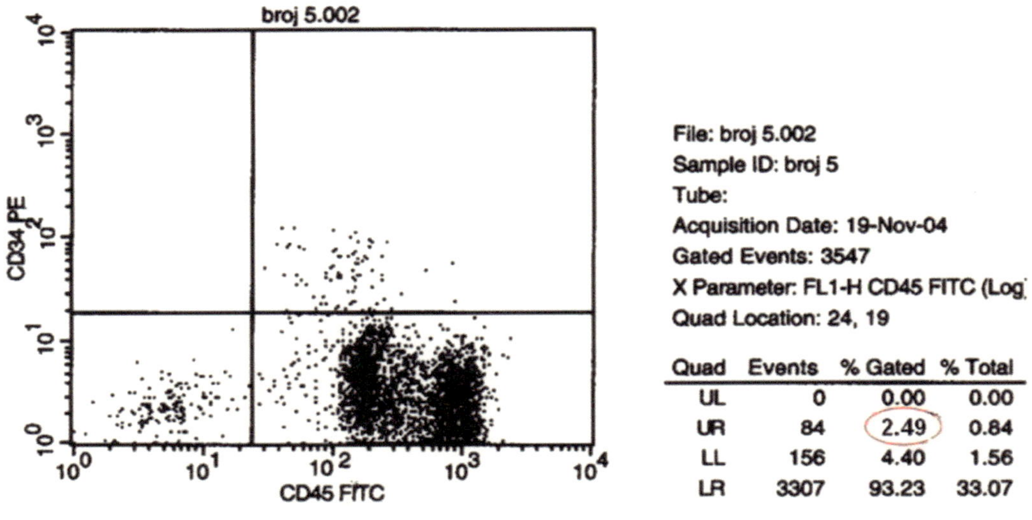

Figure 5. The ratio of the CD45/$CD34^+$ cells in peripheral blood harvest

For autologous SC transplants, the use of the best freezing process and choice of the most appropriate cryoprotective agent is required (optimized cryobiosystem). Nowadays a variety of protocols are used in blood–derived cell freezing practice. Generally, microprocessor–controlled (controlled–rate) freezing is more efficient than uncontrolled–rate (without programmed cooling) procedure due to better cell recovery. Our earlier results obtained for cryopreserved bone marrow cells and peripheral blood mononuclear cells were in agreement with these findings [8, 52]. These results imply a different "cryobiological request" of MRA cells in comparison with the mature progenitors. Moreover, our clinical studies showed that therapeutic use of the SCs – cryopreserved by our own controlled-rate system resulted with high cell recovery (91%) and rapid posttransplant hematopoietic reconstitution – on the 11th day in average [2, 8].

We have also investigated SC–harvesting protocols with optimized cell source, collection time–point and processed blood volume, CD34–threshold dose (calculed by ideal body mass), as well as immature ($CD34^+/CD33^-$, $CD34^+/CD38^-$, $CD34^+/DR^-$, $CD34^+/CD90^+$) vs. mature ($CD34^+/CD33^+$, $CD34^+/CD38^+$, $CD34^+/DR^+$, $CD34^+/CD90^-$) CD34–subset ratio in harvest [8, 26, 28, 50].

http://dx.doi.org/10.5772/54473

Several data related to our immature vs. mature CD34⁻subset investigations of cells collected from different sources are presented in Table 2.

	PB–SCs I	PB–SCs II	BM–SCs
CD34 PE / CD90 FITC [%]	1,72±1,47	1,25±0,82	2,72±2,06
CD34 PE / CD38 FITC [%]	2,02±1,18	2,02±1,18	2,3±1,16
CD34 PE / HLA–DR FITC [%]	2,01±0,92	2,0±0,92	2,0±0,88
CD34 PE / CD33 FITC [%]	1,90±1,23	1,9±1,23	2,75±1,12

PB–SCs I = stem cell collected from peripheral blood after mobilization with chemotherapy and rHuG–CSF; PB–SCs II = stem cell collected from peripheral blood after mobilization with rHuG–CSF alone; BM–SCs = stem cell collected from bone marrow.

Table 2. Ratio of CD34 cell markers using double staining

Finally, we found that the use of large volume vs. conventional (repetitive) apheresis resulted in improved $CD34^+$ yield and viability (7–AAD flow cytometric assay) [2, 8, 50]. The harvesting of higher ratio of immature vs. mature CD34–subsets correlated with BM repopulation ability (complete and long–term engraftment) and rapid hematopoietic reconstitution, as well as superior organ repair or SC regenerative potential.

SCs are considered optimal targets for gene transduction due to their ability to renew themselves and differentiate into progeny cells and generate a self-perpetuating cell population that contains the transduced gene for the lifetime of the patient. Specific diseases that could be candidates for SC gene therapy include thalassemia, sickle cell anemia, Fanconi anemia, severe combined immune deficiency secondary to adenosine deaminase deficiency or purine nucleoside phosphorylase deficiency, chronic granulomatous disease, leukocyte adhesion deficiency, Gaucher's disease, and a variety of other metabolic/storage deficiencies. UCB derived SCs potentially could be used to correct genetic deficiencies at birth after successful gene transduction and autologous transplant [2, 50].

13. Stem cells in regenerative medicine – A rapid consideration

Cardiac repair following SC application.The occurrence of heart failure following acute myocardial infarction – during the hospital stay and during the next few months or years – is high (up to 50%). Patients' mortality with heart failure after infarction is also considerable. The left ventricle dilatation occurs in even approximately one third of the patients, reperfused effectively with primary angioplasty. The incidence of heart failure after infarction has increased and mortality decreased with better reperfusion therapy [2, 3, 28]. Consequently, it is imperative to develop a curative approach to prevent of myocardium remodeling. The SC therapy is a new and promising manner of an infarcted heart healing.

It is generally accepted that adult SCs from different tissues may transdifferentiate in special situations into cardiomyocytes or endothelial cells. In addition, the presence of an important quantity of cardiac cells is confirmed in proliferative state in peri-infarction region. The first source of these "regenerative cells" is maybe cardiac SCs – which are in the inactive stage in intact (undamaged) myocardium – but following infarction they differentiate into cardiomyocytes, smooth muscle cells and endothelial cells [2, 3]. The next discovery is that for the period of infarction, myocardial ischemia initiates release of some cytokines, growth-factors and chemokines – which induce SC mobilization from other niches and their homing into the damaged myocardium. The knowledge of these processes is important because the treatment efficacy depends on artificial *ex vivo* and/or *in vivo* intensification of some "steps" in order to make regenerative process more beneficial.

The exact mechanism by which SCs create a protective effect resulting in tissue/organ repair and heart function improvement is also a matter of debate. A number of possible mechanisms have been proposed: a) SC transdifferentiation into cells of other lineages (cardiomyocytes or endothelial cells) resulting in formation of new tissue; b) mobilization of tissue specific SCs from the BM that home to the damaged tissue and participate in tissue regeneration; c) fusion of the SCs with cells of the target tissue giving rise to new cells; and d) creation of a milieu (perhaps by releasing growth factors) that enhances regeneration of endogenous cells [2, 3, 50].

At present, BM is the most frequent source of cells used for clinical cardiac repair. It contains a complex mixture of progenitor cells – including SCs; so-called side population (SP) cells, which account for most if not all long-term self-renewal and reconstitute the full panoply of hematopoietic lineages after single-cell grafting; a subset of mesenchymal or stromal cells (MSCs), which are already defined; multipotent adult non-hematopoietic progenitor cells (MAPCs – for example, VSEL cells), which can differentiate into all possible lineages, and a fraction of TCSC discovered recently by Ratajczak et coworkers [2, 38–44]. These TCSCs circulate at the highest level and thus accumulate in BM during rapid body growth and become a reserve pool of SCs for tissue/organ regeneration. They are chemo-attracted from PB to injured organs by signaling proteins, such as stromal cell-derived factor-1, which become highly expressed in damaged heart tissue. For therapeutic purposes, marrow is aspirated, the entire MNC fraction is obtained – a mixed combination of mentioned cells – or specific subpopulations are purified and isolated cells are injected into the heart without need of further *ex vivo* manipulation/expansion.

In conclusion, current challenges for cell-based therapy in cardiac repair include identifying the origins of the novel cardiac SCs found inside heart, pinpointing biologically active cells from BM and other cell populations, optimizing cell mobilization and homing, increase of survival of grafted SCs, and exploiting cell therapy as a platform for secretor signals. Thus, we need a lot of basic research and randomized clinical trials to define the exact role of this probably revolutionary therapy for ischemic heart disease.

Application of SCs for liver and pancreas regeneration. The growing donor organ shortage requires consideration of alternative emerging technologies. Regenerative medicine may offer novel strategies to treat patients with end-stage or severe organ failure.

http://dx.doi.org/10.5772/54473

The final purpose of SC therapy, organ repopulation strategies and tissue engineering is to regenerate tissues/organs or to produce new grafts/organs for transplant. With the expansion of complete organ "decellularization" methods the equation of organ shortage could be radically altered in the future. Decellularized organs provide the ideal transplantable platform with all the essential microstructure and extra-cellular signals for cell connection (homing), transdifferentiation, tissue vascularization, etc. Novel systems to re-engineer organs may have key connotations for the fields of regenerative biology and ultimately organ transplant [2, 50].

Currently available β-cell replacement therapies for patients with Type 1 diabetes (T1D), including islet and pancreas transplant, are largely successful in restoring normal glucose metabolism. However, there are data concerning the use of SCs to generate β-cells for islet transplant, indicating the need for improved protocols for their derivation and full maturation. Researchers also considered evidence indicating that adult SCs may affect islet transplant by improving the viability of engrafted islets and controlling immune-related reactions to islet antigens. A novel SC-based applications or regeneration-type approaches include stimulation of endogenous regenerative mechanisms or inducing reprogramming of non-β cells into β cells. Because these strategies would finally generate allogeneic or syngeneic β cells, the control of alloimmunity or autoimmunity in addition to replacing lost β cells will be of the greatest importance [2, 50].

For the SC treatment of our patients with liver failure (n = 8) and T1D (n = 4), cells were harvested from PB following mobilization (rHuG-CSF 10μg/day; 5 days). The mean volume of processed blood was 15.2±1.6 L (ratio: 12.8 – 18.4 L). The total count of MNC and $CD34^+$ collected cells were $6.4\pm3.1x0^9$ and $1.6\pm08x10^7$, respectively. Cells were applied after immunomagnetic selection and *ex vivo* transdifferentiation and expansion across catheter [2, 50].

The use of SCs in neurology/neurosurgery. Alzheimer's, Parkinson's and Huntington's diseases, amyotrophic lateral sclerosis (ALS), and Friedreich's ataxia are the most common human neurodegenerative diseases – pathologically characterized by a progressive and specific loss of certain neuronal populations. Currently there are no effective clinical therapies for many of these diseases. The recently acquired ability to reprogram human adult somatic cells to "induced pluripotent SCs" (iPSCs) in culture may provide a powerful tool for *in vitro* neurodegenerative disease modeling and an unlimited source for cell replacement therapy. Reprogramming of somatic cells into iPSCs ushered in a new era of regenerative medicine. Human iPSCs give potent new approaches for disease modeling, drug testing, developmental studies, and therapeutic applications.

We earlier largely described the specific therapeutic actions and clinical use of SCs in neurology/neurosurgery [2]. Cerebral tumors, stroke, neurodegenerative diseases, brain and spinal injuries are used as examples with different limitations and possibilities to be approached with adult SC and/or different regenerative treatments. It is clear, that despite a spectrum of successful approaches, there are current limitations in this field of therapeutic interventions, which makes the research more intriguing and opens the new avenues for the development of novel concepts, their future prove, and possible application.

Finally, in our center SC are applied the first time in the treatment of ALS female patients (intratecal application of non-manipulated BM derived cells; two repeated treatments) and another female patient after brain infarction (intra-arterial cell injection using percutaneous catheter). Logically, preliminary or particularly definitive conclusions can only be drawn from larger, randomized, controlled clinical trials.

14. The treatment of large myocardial infarction by intracoronary applied rHuG-CSF facilitated BM derived SCs – Our experience

14.1. Introduction

Intracoronary autologous, BM derived SC transplant for the treatment of myocardial infarction went through the three important steps during the last decade. In the first step, small non-randomized trials of Strauer [53] and TOPCARE study [54], established the basic methodology of SC harvesting, cell processing and intracoronary delivery and confirmed the safety and regenerative potential of SCs for the improvement of myocardial viability and function after infarction.

The next step representing two landmark studies has brought controversial results. The REPAIR-AMI [55] has showed that intracoronary delivery of BM derived SCs led to the improvement of six months global ejection fraction and lowers the major cardiac adverse events. On the contrary, the ASTAMI trial [56] has failed to prove any benefit of early intra-coronary application of BM derived SCs to the global ejection fraction and left ventricle remodeling measured by magnetic resonance imaging at baseline and after four months of anterior infarction. However, those studies suggested several important issues. The first of all were that choice of patients with more severe damaged myocardium and the delayed SC delivery for at least five days after infarction resulted to better results of SC regenerative capacity. The second key conclusion is that SC processing process might be essential for successful SC therapy after myocardial infarction.

In the third step, two relatively large randomized studies tried to define the efficacy of different SC population for treatment of myocardial infarction. The REGENT study [57] used selection of $CD34^+/CD\text{-}CXC4^+$ cells, and the HEBE trial [58] examine selection of PB derived MNCs and compare the results with two control groups, non-selected BM origin MNCs and controls. These trials didn't show any usefulness of cell selection and suggested again that patients with more damaged myocardium had better improvement with intracoronary SC delivery. On the other hand, our study has showed that there is a limit for the amount of myocardial necrosis in which we can achieve improvement of myocardial function after intracoronary SC delivery and the patients with the huge loss of myocardium has no any benefit from the cell therapy [26–28]. However, improvement of global and regional left ventricle function was modest and faraway from the expected in all studies. The next steps in SC therapy are *ex vivo* expansion of the number and regenerative capacity of harvested SCs, *in vivo* facilitation of that, and improvement of methods of SC delivery.

http://dx.doi.org/10.5772/54473

In our Center for regenerative medicine, total of 60 patients were treated by non-manipulated or *ex vivo* cytokine stimulated BM derived cells (collected in steady state hematopoiesis or after priming of the marrow). Cells were applied across percutaneous catheter intracoronarly or directly into the myocardium (transpericardial approach). In the most recent stage, we investigated the effects of rHuG-CSF facilitated BM (primed BM) SC therapy on improvement of the global and regional function of left ventricle after large myocardial infarction.

14.2. Methodology

The main inclusion criteria for the enrollment in the study are: patients with the first ST segment elevation myocardial infarction; age younger than 71 years old; successful percutaneous coronary intervention on the infarction artery inside the 24 hours from the onset of pain and with the left ventricle ejection fraction lower than 41% at the fifth day estimated by the transthoracic echocardiography. The main exclusion criteria are the presence of other serious illness, any pre-infarction significant damage of the heart, allergy to aspirin and resistance to clopidogrel, and the presence of the symptoms and signs of heart failure five days after infarction. The local Ethical Committee approved the study and all patients were given written informed consent.

Pre-transplant examination. At the fifth day of infarction, global and regional left ventricle systolic function together with the end-diastolic and end-systolic volumes are measured by the transthoracic echocardiography. Infarction size is estimated by the Technetium-sestamibi myocardial scintigraphy between the 5-8 days from the infarction.

Stem cell harvesting and application. The day before BM harvest, patients receive 5 – 10 μg/kgbm of rHuG-CSF. Between 7 – 12 days from the infarction in the general anesthesia 300 mL of the BM is harvested from the posterior iliac crests. After that BM was filtered and processed to the final volume of 30 – 50 mL of concentrated mono-nuclear cell suspension. Boluses of 10 mL are injected through the diagnostic catheter into the infarction related coronary artery. Patients with rHuG-CSF facilitated SC therapy received $18.4 x 10^8$ of MNCs and patients without rHuG-CSF received $7.9 x 10^8$ of the MNCs.

Control groups. There are two control groups with the same inclusion and exclusion criteria. The first are patients who did not submit to any SC procedure. And the second represents the patients who were treated with the autologous intracoronary, BM derived SC therapy without rHuG-CSF.

The follow-up. Clinical examination is scheduled for the one, fourth and sixth months after infarction and every 6 months after that. Echocardiography measurement of global and left ventricle ejection fraction and volumes is planned after 4 months and every year after infarction. Myocardial scintigraphy is planned after 4 months and after two years from the infarction.

End points. The main end points are comparison of the 4 months and 2 years change in left ventricle ejection fraction, volumes and infarction size between three groups.

14.3. Results

The baseline characteristics of three groups of patients were similar except that control group were slightly older than patients in both SC groups and have more often multivessel disease. Gender and risk factors distribution were similar between groups. There is also no difference in ischemic time in three groups. Baseline global left ventricle ejection fraction and infarction size were similar in all three groups. However, both end-diastolic and end-systolic cardiac indices were lower in patients with rHuG-CSF facilitated SC therapy.

After 4 months left ventricle ejection fraction has improved in group of patients treated with SCs and in control group but did not reach statistical significance in the group treated with rHuG-CSF facilitated SC therapy because of small number of patients in that group (Table 1). Infarction size has the same pattern (Table 3). End-diastolic and end-systolic volumes increased in all groups but also did not get to significant statistical difference in the rHuG-CSF facilitated SC group.

Parameters	SC therapy in AMI n=19	G-CSF facilitated SC therapy in AMI n=5	Control group n=17	p
Infarction size at baseline LV%±SD	28,4±11,3	35,6±8,0	31,4±12,8	ns
Infarction size after 6 months LV%±SD	25.2±12.6	25.2±8.6	27.9±10.7	ns
p	0.001	0.068	0.001	
LVEF at baseline % ± SD	32,9±4,1	36,4±3,0	34,3±5,2	ns
LVEF after 4 months % ± SD	37,0±9,0	43,8±3,0	36,9±8,2	0,01
p	0.004	0.313	0.004	
EDVCI at baseline ml/m^2 ± SD	68,3±11,3	46,1±10,0	67,8±17,6	0,01
EDVCI after 4 months ml/m^2 ± SD	75,7±15,7	54,3±6,1	73,8±20,4	0,053
p	0.024	0.161	0.004	
ESVCI at baseline ml/m^2 ± SD	44,8±9,8	28,0±4,4	45,0±15,1	0.02
ESVCI at 4 months ml/m^2 ± SD	47,8±14,3	30,3±2,6	47,5±17,6	0,07
p	0.001	0.142	0.002	

Table 3. Infarction size end left ventricle systolic function and volumes at baseline and after 4 months.

http://dx.doi.org/10.5772/54473

Difference between baseline and 4-months infarction size is significant only in patients with rHuG-CSF facilitated SC therapy (Figure 6). There was no significant difference between the change of LVEF at baseline and after 4 months (Figure 6).

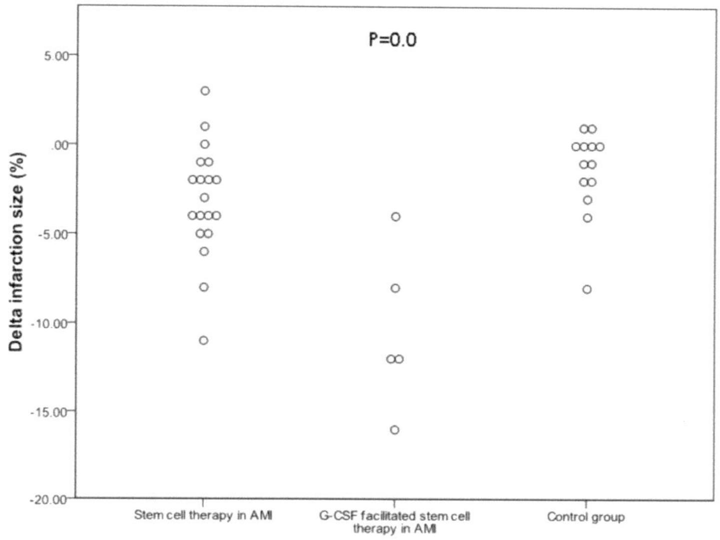

Figure 6. Change in infarction size between baseline and 4-months among three groups of patients

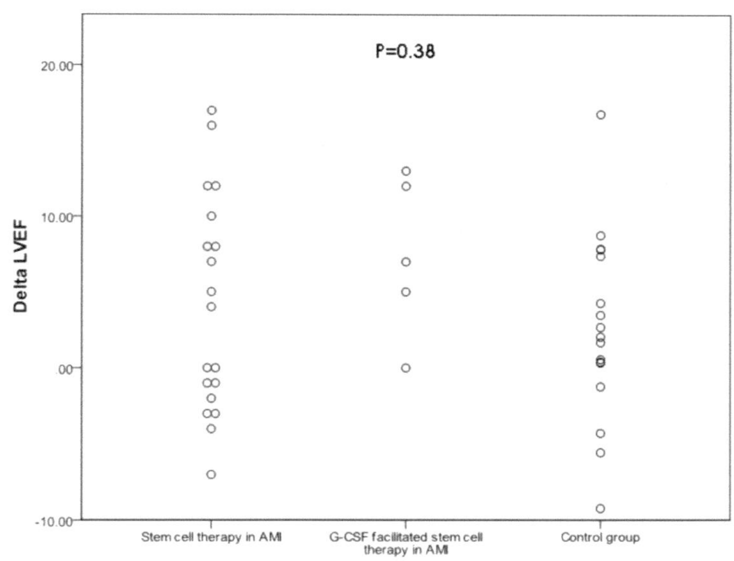

Figure 7. Change in left ventricle ejection fraction: baseline vs. 4-months among three groups of patients

15. Conclusion

Our preliminary results have shown that rHuG-CSF facilitated autologous, intracoronary SC transfer was safe; between two-three times higher number of MNCs were given and there was a trend toward larger increase of 4-months ejection fraction and greater decrease of the infarction size than the control groups. Any procedure that increases the left ventricle ejection fraction for more than 5% after a several months follow-up would be of great clinical and economic value. Autologous BM derived, intracoronary SC transfer in the second week of large myocardial infarction very probably improved the global left ventricle ejection fraction by 3 – 5% although results of the published trial are controversial. Granulocyte colony stimulating factor given alone for several days after myocardial infarction did not improve significantly global ejection fraction in the several trials, but it seems that its early application in patients with larger infarction could be useful. Further investigation is needed for the justification of rHuG-CSF facilitated, BM derived SC therapy in the early phase of acute ST elevation myocardial infarction.

Author details

Bela Balint[1,2,3*], Slobodan Obradovic[3,4], Milena Todorovic[5,6], Mirjana Pavlovic[7] and Biljana Mihaljevic[5,6]

*Address all correspondence to: beciusb@yahoo.com

1 Institute for Medical Research, University of Belgrade, Serbia

2 Institute for Transfusiology and Hemobiology of MMA, Belgrade, Serbia

3 Faculty of Medicine MMA, University of Defense, Serbia

4 Clinic of Emergency Medicine of MMA, Belgrade, Serbia

5 Clinic for Hematology, Clinical Center of Serbia, Belgrade, Serbia

6 Faculty of Medicine, University of Belgrade, Serbia

7 Department of Computer and Electrical Engineering and Computer Science, FAU, Boca Raton, FL, USA

References

[1] Mayorga M, Finan A, Penn M. Pre–transplantation specification of stem cells to cardiac lineage for regeneration of cardiac tissue. Stem Cell Rev Rep 2009; 5 (1): 51–60.

[2] Pavlovic M, Balint B. Stem Cells and Tissue Engineering. New York: Springer. In press 2012.

[3] Obradovic S, Balint B, Trifunovic Z. Stem Cell Therapy in Myocardial Infarction Clinical Point of View and the Results of the REANIMA Study. In: Gholamrezanezhad A. (Ed). Stem cells in clinic and research. Rijeka: InTech; 2011. p. 233–58.

[4] Janssens S, Dubois C, Bogaert J, Theunissen K, Deroose C, Desmet W, et al. Autologous bone marrow-derived stem-cell transfer in patient with ST-segment elevation myocardial infarction: double-blind, randomized controlled trial. Lancet 2006; 367: 113–21.

[5] Hirsch A, Nijveldt R, Vleuten PA, Tijssen J, Giessen W, Tio R, et al. Intracoronary infusion of mononuclear cells from bone marrow or peripheral blood compared with standard therapy in patients after acute myocardial infarction treated by primary percutaneous coronary intervention: results of the randomized controlled HEBE trial. Eur Heart J. 2011; 32 (4): 1736–47.

[6] Balint B, Ivanovic Z. Mobilized peripheral blood stem cell transplants. Bilt Transfusiol 1995; 24: 5–11.

[7] Moog R. Apheresis techniques for collection of peripheral blood progenitor cells. Transfus Apheresis Sci. 2004; 31 (3): 207–20. XXX

[8] Balint B, Ljubenov M, Stamatović D, Todorović M, Pavlović M, Ostojić G, et al. Stem cell harvesting protocol research in autologous transplantation setting: large volume vs. conventional cytapheresis Vojnosanit Preg 2008; 65 (7): 545–51.

[9] Gluckman, E, Broxmayer HE, Auerbach AD. Hematopoetic reconstruction in a patient with Fanconi anemia by means of umbilical cord from HLA identical sibling. N Engl J Med 1989; 321: 1174–8.

[10] Hiett AK, Britton KA, Hague NL, Brown HL, Stehman FB, Broxmeyer HE. Comparison of hematopoietic progenitor cells in human umbilical cord blood collected from neonatal infants who are small and appropriate for gestational age. Transfusion 1995; 35: 587–91. XXX

[11] Engelfriet CP, Reesink HW. Use of cord blood progenitor cells as an alternative for bone marrow transplantation. Vox Sang 1998; 75: 156–72.

[12] Skoric D, Balint B, Petakov M, Sindjic M, Rodic P. Collection strategies and cryopreservation of umbilical cord blood. Transfusion Medicine 2007; 17 (2): 107–13.

[13] Thomas ED, Lochte HLJr, Lu WC, Ferrebee JW. Intravenous infusion of bone marrow in patients receiving radiation and chemotherapy. N Engl J Med 1957; 257: 491–496.

[14] Thomas ED, Lochte HLJr, Cannon JH, Sahler OD, Ferrebee JW. Supralethal whole body irradiation and isologous marrow transplantation in man. J Clin Invest 1959; 38: 1709–16.

[15] Mathe G, Jammet H. Transfusions et greffes de moelle osseusse homologue chez des humaines irradies a haute danse accidentellement. Rev Fr Etud Clin Biol 1959; 4: 226–38.

[16] Andrews GA. Criticality accidents in Vinca, Yugoslavia and Oak Ridge Tennessee. Am J Roentgenol Radium Ther Nucl Med 1965; 93: 56–74.

[17] Dausset J. Iso-leuco-anticorps. Acta Haemutol 1958; 20: 156.

[18] Thomas ED et al. Allogeneic marrow grafting for hematologic malignancy using HL–A matched donor–recipient sibling pairs. Blood 1971; 38: 267–87.

[19] Lasky LC, Ash RC, Kersey JH. Collection of pluripotential hematopoietic cells by cytapheresis. Blood 1982; 59: 822–7.

[20] Brecher ME, editor. Technical Manual, 15th edition. Bethesda: AABB Press; 2005.

[21] Strauer BE, Brehm M, Zeus T, Köstering M, Hernandez A, Sorg RV. Repair of infarcted myocardium by autologous intracoronary mononuclear bone marrow cell transplantation in humans. Circulation 2002; 106 (5): 1913–8.

[22] Hassink RJ, de la Rivere AB, Mummery CL, Doevendans PA. Transplantation of cells for cardiac repair. J Am Coll Cardiol 2003; 41 (5): 771–7.

[23] Obradovic S, Rusovic S, Balint B, Ristic–Andjelkov A, Romanovic R, Baskot B et al. Autologous bone marrow–derived progenitor cell transplantation for myocardial regeneration after acute infarction. Vojnosanit Pregl 2004; 61 (5): 519–29.

[24] Flynn A, O'Brien T. Stem cell therapy for cardiac disease. Expert Opin Biol Ther 2011; 11 (2): 177–87.

[25] Wu J, Li J, Zhang N, Zhang C. Stem cell–based therapies in ischemic heart diseases: a focus on aspects of microcirculation and inflammation. Basic Res Cardiol 2011; 106 (3): 317–24.

[26] Balint B, Todorovic M, Jevtic M, Ostojic G, Ristanovic E, Vojvodic D, et al. The use of stem cells for marrow repopulation and in the field of regenerative medicine. Maked Med Pregl 2009; 63 (Suppl 7 5): 12–7.

[27] Balint B. Stem cells – unselected or selected, unfrozen or cryopreserved: marrow repopulation capacity and plasticity potential in experimental and clinical settings. Maked Med Pregl 2004; 58: 22–4

[28] Obradovic S, Balint B, Romanovic R, Trifunovic Z, Rusovic S, Baskot B, et al. Influence of intracoronary injections of bone–marrow–derived mononuclear cells on large myocardial infarction outcome: quantum of initial necrosis is the key. Vojnosanit Pregl 2009; 66 (2): 998–1004.

[29] Lunde K, Solheim S, Aakhus S, Arnesen H, Abdelnoor M, Egeland T, et al. Intracoronary injection of mononuclear bone marrow cells in acute myocardial infarction. N Engl J Med 2006; 355: 1199–209.

[30] Bajek A, Olkowska J, Drewa T. Mesenchymal stem cells as a therapeutic tool in tissue and organ regeneration. Postepy Hig Med Dosw 2011; 65: 124–32.

[31] Vlaski M, Lafarge X, ChevaleyreJ, Duchez P, Boirona J–M, and Ivanovic Z: Low oxygen concentration as a general physiologic regulator of erythropoiesis beyond the EPO–related downstreamtuning and a tool for the optimization of red blood cell production ex vivo Exp Hematol, 2009; 37: 573–84.

[32] De Copi P, Bartsch G Jr, Minhaj Ms, Atala A et al: Isolation of amniotic stem cell lines with potential for therapy. Nature Biotechnology 2007; 25: 100–6.

[33] Kim J, Lee y, Hwanf KJ, Kwon HC, Kim SK, Cho DJ, Kang SG, You J. Human amniotic fluid–derived stem cells have characteristics of multipotent stem cells. Cell prolif 2007; 40 (1): 75–90.

[34] Jendelova P., Herynek V, Urdzikova L, Glogarova K, Kroupova J, Anderson B, Vryja V, Burtain M, Hajek M, and Sykova J. 2004. Magnetic resonance tracking of transplanted Bone Marrow and Embryonic Stem cells labeled by Iron Oxide Nanoparticles in Rat Brain and Spinal Cord. Jour Neurosci Res 76: 232–43.

[35] Marshall CT, Lu C, Winstead W, Zhanf X, Xiao M, Harding G, Klueber KM, and Roisen FJ. The therapeutic potential of human olfactory–derived stem cells. Histol Histopatol 2006, 21: 633–643

[36] Lindwall O, Kokaia Z, Martinez–Serrano A. SCs for the treatment of neurological disorders. Nature 2006; 441: 1094–6.

[37] D'Ippolito G, Sylma D, Guy A. Howadr–, Philippe M, Roos BA, Schiller PC. Marrow–isolated adult multilineage inducible (MIAMI) cells, a unique population of postnatal young and old human cells with extensive expansion and differentiation potential. J Cell Science 2004; 117: 2971–81.

[38] Ratajzcak MZ, Kucia M, Reca R, Majka M et al, Stem cell plasticity revised: CXR4 positive cells expressing mRNA for earlu muscle, liver and neural cells "hide out" in the bone marrow. Leukemia 2004; 19 (1): 29–40

[39] Serafini M, Dylla SJ, Oki M, Heremans Y, Tolar J, Jiang Y, et al. Hematopoietic reconstitution by multipotent adult progenitor cells: precursors to long–term hematopoietic stem cells. J Exp Med 2007 Jan 22; 204 (1): 129–39

[40] Kucia M, Wojakowski W, Ryan R, Machalinski B, Gozdzik J, Majka M, Baran J, Ratajczak J, and Ratjczak MZ. The migration of bone marrow–derived non–hematopoietic tissue–commited stem cells is regulated in and SDF-1-, HGF-, and LIF dependent manner. Arch Immunol Ther Exp 2006, 54, 2: 121–135

[41] Kucia M, Zhang YP, Reac R, Wysoczynski M, Machalinski B, Majka M, Ildstad ST, Ratajczak JU, Chields CB, and Ratajczak MZ. Cells enriched in markers of neural tissue–commited stem cells reside in the bone marrow and are mobilized into the peripheral blood following stroke. Leukemia 2006, 20: 18–28

[42] Kucia M, Reca R, Campbell FR, Surma–Zuba E, Majka M, Ratajczak M, and Ratajczak MZ A population of very small embryonic – like (VSEL) CXR4+SSEA–1+Oct4+ stem cells identified in adult bone marrow. Leukemia 2006, 20: 857–69

[43] Ratajczak MZ, Zuba–Surma E, Kucia M, Poniewierska A, Suszynska M, Ratajczak J. Pluripotent and multipotent stem cells in aduld tissue. Adv Med Sci 2012; 19: 1–17.

[44] Ratajczak MZ, Suszynska M, Pedziwiatr D, Mierzejewska K, Greco NJ. Umbilical cord blood–derived very small embryonic like stem cells (VSELs) as a source of pluripotent stem cells for regenerative medicine. Pediatr Endocrinol Rev 2012; 9: 639–43.

[45] Arthur A, Rychkov G, Shi S, Koblar SA, GronthosS. Adult human dental pulp stem cellsdifferentiate toward functionally active neuronsunder appropriateenvironmental cues. Stem Cells 2008; 26: 1787–95

[46] Todorovic V, Markovic D, Milosevic–Jovcic N, Petakov M, Balint B, Colic M. Dental pulp stem cells – potential significancein regenerative medicine. Stom Glas 2008; 55: 170–9.

[47] Morsczeck K, Frerich B, Driemel O. Dental Stem Cell Patents. Clin Oral Invest 2008; 12: 113–118.

[48] Balint B. Coexistent cryopreservation strategies: microprocessor–restricted vs. uncontrolled–rate freezing of the "blood–derived" progenitors/cells. Blood Banking Transf Med 2004; 2 (2): 62–71.

[49] Balint B. Stem and progenitor cell harvesting, extracorporeal "graft engineering" and clinical use – initial expansion vs. current dillemas. Clin Appl Immunol 2006; 5 (1): 518–27.

[50] Balint B, Todorovic M, Jocic M, Stamatovic D. Haemo-therapeutic approach to cell and organ transplants. Belgrade: ART-Press; 2011.

[51] Balint B, Stamatovic D, Todorovic M, Elez M, Vojvodic D, Pavlovic M, Cucuz–Jokic M. Autologous transplant in aplastic anemia: quantity of $CD34^{+}/CD90^{+}$ subset as the predictor of clinical outcome. Transf Apher Sci 2011; 45 (2): 137–41.

[52] Balint B, Ivanovic Z, Petakov M, Taseski J, Jovcic G, Stojanovic N, et al. The cryopreservation protocol optimal for progenitor recovery is not optimal for preservation of MRA. Bone Marrow Transpl 1999; 23: 613–9.

[53] Strauer BE., Brehm M., Zeus T., Köstering M., Hernandez A., Sorg RV, et al. (). Repair of infracted myocardium by autologous intracoronary mononuclear bone marrow cell transplantation in humans. Circulation 2002; 106: 1913–8.

[54] Assmus B., Rolf A., Erbs S., Elsässer A., Haberbosch W., Hambrecht R., et al. Clinical outcome 2 years after intracoronary administration of bone marrow-derived progenitor cells in acute myocardial infarction. Circulation Heart Failure 2010; 3: 89–96.

[55] Schächinger V, Erbs S, Elsässer A, Haberbosch W, Hambrecht R, Hölschermann H, et al. Improved clinical outcome after intracoronary administration of bone-marrow-

derived progenitor cells in acute myocardial infarction: final 1-year results of the REPAIR-AMI trial. Eur Heart J 2006; 27: 2775–83.

[56] Lunde K, Solheim S, Aakhus S, Arnesen H, Abdelnoor M, Egeland T, et al. Intracoronary injection of mononuclear bone marrow cells in acute myocardial infarction. N Engl J Med 2006; 355: 1199–209.

[57] Tendera M, Wojakowski W, Ruzytto W, Chojnowska L, Kepka C, Tracz W, et al. Intracoronary infusion of bone-marrow derived selected CD34+CXCR4+ cells and non-selected MNCs in patients with acute STEMI and reduced left ventricular ejection fraction: results of randomized, multicentre Myocardial Regeneration by Intracoronary Infusion of Selected Population of STEM Cells in Acute Myocardial Infarction (REGENT) Trial. Eur Heart J 2009; 30: 1313–21.

[58] Hirsch A., Nijveldt R., Vleuten P.A., Tijssen J., Giessen W., Tio R. et al. Intracoronary infusion of MNCs from bone marrow or peripheral blood compared with standard therapy in patients after acute myocardial infarction treated by primary percutaneous coronary intervention: results of the randomized controlled HEBE trial. Eur Heart J 2010

Chapter 8

Hematopoietic Stem Cells in Chronic Myeloid Leukemia

Antonieta Chávez-González,
Sócrates Avilés-Vázquez,
Dafne Moreno-Lorenzana and Héctor Mayani

Additional information is available at the end of the chapter

http://dx.doi.org/10.5772/54651

1. Introduction

Chronic Myeloid Leukemia (CML) is a clonal disease, originated at the level of Hematopoietic Stem Cells (HSC) and characterized by the presence of the *Philadelphia* (Ph) chromosome and its oncogenic product p210Bcr-Abl. Such a protein has been shown to be essential for malignant transformation, since it is capable of altering cell adhesion, proliferation and apoptosis.

Current treatment options in CML include tyrosine kinase inhibitors (Imatinib, Nilotinib and Dasatinib), compounds that inhibit the activity of the BCR-ABL protein. However some patients will develop resistance or intolerance to these drugs and resistance has been associated with different mechanism including the quiescence of leukemic stem cells and Pgp or Src kinase overexpression.

In this chapter we focus on the basic biology of hematopoietic stem and progenitor cells from CML and analyze the most relevant and current concepts in this area.

2. Chronic myeloid leukemia

Chronic myeloid leukemia (CML) is a lethal hematological malignancy characterized by the abnormal amplification of the myeloid (mainly granulocityc) compartment of the hematopoietic system. It originates from the transformation of a primitive hematopoietic cell that suffers a t(9;22) (q34; q11) balanced reciprocal translocation that results in the generation of the Philadelphia chromosome (Ph). Ph produces BCR-ABL, a constitutively active tyrosine kinase that drives a wide variety of physiological alterations [1].

CML was initially described in 1845 by John Hughes Bennett, who reported the case of a patient with "milky" blood and suggested that it was an infectious disease that caused hypertrophy of the liver and spleen, leading to the patient's death. A few weeks later, Rudolf Virchow reported a similar case, but, in contrast to Bennett, he suggested that the disease was not infectious and implied an increase in the number of blood cells. He coined the term leukemia (from the Greek *leukos*, white, and "Aemia", blood). In 1870, Neumann described that leukemia cells originate in the bone marrow; almost one hundred years later, in 1960, Nowel and Hungerford reported that in all cases of this malignancy there was a small, abnormal chromosome 22. However, was until 1973 that Janet Rowley described that the abnormal chromosome was caused by a reciprocal translocation between the long arms of chromosomes 9 and 22, designating the name of Philadelphia (Ph) chromosome[2, 3].

2.1. Epidemiology and clinical characteristics

Chronic myelogenous leukemia has a worldwide incidence of 1-2 cases per 100,000 individuals [4]. The average age at diagnosis is 60 years; it occurs less frequently in young people and a tendency to increase exponentially with age has been observed. There is no geographic or genetic predisposition to acquire this condition, although some authors have associated it with exposure to high doses of ionizing radiation. The current CML prevalence of 24,000 affected patients in the United Sates is relatively low; it is expected to increase significantly over the next 20 years as a result of widespread use of BCR-ABL tyrosine kinase inhibitor therapy [5]. In Mexico, there are no official data on the incidence of such a disease, however, it has been estimated that there are about 80,000 cases of leukemia and 10% corresponds to CML [6].

The clinical presentation often includes granulocytosis, spenomegaly and marrow hypercellularity; however about 40% of patients are asymptomatic and their diagnosis is based on abnormal blood cell counts [1]. The natural course of the disease involves three sequential phases, namely chronic, accelerated and blast crises. Ninety percent of patients are diagnosed in chronic phase and they remain in it for 3 to 8 years. In this phase, the blood cells retain their ability to differentiate until the illness progresses to the accelerated phase, which is characterized by the egress of immature cells into the bloodstream. Finally, the disease progresses to the blast crisis, defined by the presence of 30 percent or more leukemic cells in peripheral blood or marrow or extramedullary infiltrates of blast. During this phase the survival of patients is reduced to months and even weeks [7].

2.2. Molecular events (Bcr-Abl oncogene)

As mentioned before, the Philadelphia chromosome, which defines CML, is a shortened chromosome 22 originated from the reciprocal translocation between the long arms of chromosomes 9 and 22 [t (9; 22)] and involves addition of 3' segments of the *abl* gene (9q34) to 5' segments of the *bcr* gene (22q11) given rise to a *bcr-abl* fusion gene that transcribes a chimeric mRNA of 8.5 kb that, in turn, gives rise to a BCR-ABL fusion protein [7]. t(9;22) is evident in more than 95% of CML patients; between 5% and 10% of CML patients also present complex rearrangements that may involve one or more chromosomes in addition to 9 and 22 [8].

http://dx.doi.org/10.5772/54651

The normal human ABL gene encodes for a non-receptor tyrosine kinase that is ubiquitously expressed. Such a 145 kDa protein is involved in the regulation of the cell cycle, the response to genotoxic stress, and intracellular signaling mediated by the integrin family [9]. There are three isoforms of the BCR-ABL fusion protein all of which encode the same portion of the ABL tyrosine kinase, but differ in the length of the BCR sequence at the N-terminus. p185/p190 BCR-ABL is expressed in Acute Lypmphoblastic Leukaemia (ALL), p210 BCR-ABL is characteristic of Chronic Myeloid Leukemia, and p230 BCR-ABL has been associated with a subgroup of CML patients with a more indolent disease (Figure 1) [4].

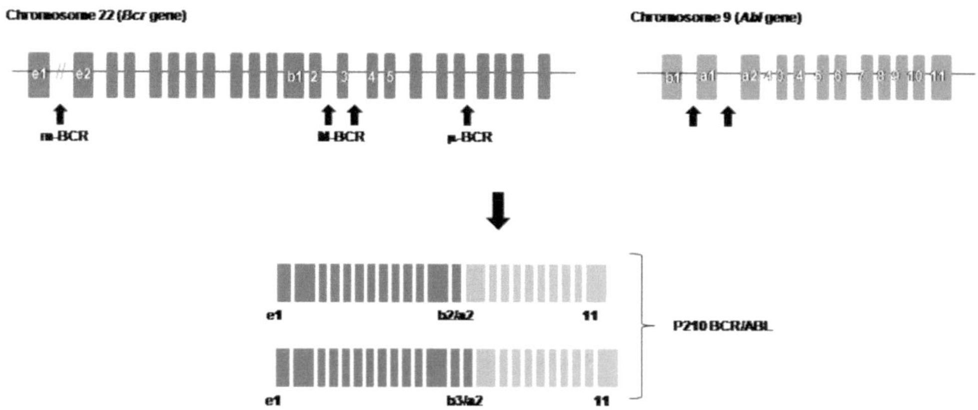

Figure 1. Structure of the Bcr-Abl gene. It is formed by a reciprocal translocation between chromosomes 22 (Bcr gene) and 9 (Abl gene). Ther M-BCR breakpoint resulting in a P210 BCR/ABL fusion transcripts b2a2 or b3a2 and they encode a protein of 210 kDa (BCR-ABLp210) present in almost all patients with Chronic Mieloid Leukemia (modified to [9]).

BCR-ABL fusion protein localizes in the cytoplasm and shows an increased and constitutive tyrosine kinase activity as a result of oligomerization of its coiled region and deletion of the SH domain of ABL. It activates a number of cytoplasmic and nuclear signal-transduction pathways involved in cell adherence, migration, inhibition of apoptosis, and induction of cell proliferation through activation of signaling proteins such as p21RAS, c-Myc, lipid kinasse PI3k, MAPk (mitogen-activated protein kinase family), tyrosine phosphatases, and signal transducer and activator of transcription (STATs) factors [9, 10].

2.3. Leukemic Stem Cells in chronic myeloid leukemia

There is an increasing body of evidence indicating that, similar to normal hematopoiesis, a quiescent stem cell population -within the CD34$^+$ cell compartment- exists in the bone marrow of CML patients. Such Leukemic Stem Cells (LSC) seem to be the ones driving CML progression, following a similar pattern to the one observed in normal hematopoiesis. That is to say, LSC give rise to CML progenitor cells, which, in turn, give rise to more mature cells.

Just like normal hematopoietic stem cells (HSC), CML stem cells express high levels of CD34, and lack the cell surface markers CD38, CD45RA, or CD71, as well as lineage-specific markers.

However, LSC are Ph$^+$/BCR-ABL$^+$, which is not present in their normal counterparts. Interestingly, it has recently been shown that a novel population of lineage-negative, CD34-negative hematopoietic stem cells from CML patients also correspond to BCR-ABL$^+$ leukemic stem cells capable to engraft immunodeficient mice [13]. Thus, it seems that most LSC are CD34$^+$ but a subpopulation may be CD34$^-$. Importantly, despite the predominance of LSC in CML, a residual population of normal hematopoietic stem cells (BCR-ABL$^-$ CD34$^+$) persists in the marrow's patient, which seems to be responsible for hematopoietic recovery after a successful treatment using Tyrosine Kinase Inhibitors (TKIs).

As mentioned before, LSC are in a quiescent state, however, they can spontaneously exit G_0 to enter a proliferating state and are capable of engrafting inmmunodeficient mice [11]. In this regard, several studies have shown that TKIs, like Imatinib, Nilotinib, Dasatinib, Bosutinib, and Lonafarnib, have antiproliferative or apoptotic effects in almost all dividing CML cells; however, the population of stem cells remains viable in a quiescent state [16-21].

In vitro studies indicate that LSCs are capable of surviving for several weeks in the absence of added growth factors due to autocrine mechanisms involving production of granulocyte colony-stimulating factor (G-CSF) and Interleukin 3 (IL-3) [12]. This, in fact, is an important difference between normal and CML HSC, since the former depends on the presence of exogenous cytokines for their growth, whereas the latter, as just mentioned, can utilize autocrine mechanisms. Although there is strong evidence that Bcr-Abl is sufficient to induce CML-like disease in transduction and transgenic murine models [14], it is still unclear whether Bcr-Abl is always the first hit in CML, since in some patients with a complete cytogenetic response after treatment, BCR-ABL transcripts are still detectable by RT-PCR, which indicates that leukemic cells persist even when the disease is reduced below detectable limits [15].

3. Functional characteristic of leukemic stem cells in CML

3.1. Proliferation

Proliferation of leukemic stem and progenitor cells is regulated by Bcr-Abl. Such a tyrosine kinase activates the Ras/Raf/MEK/ERK and JAK/STAT signal transduction pathways, and this results in an amplified proliferative state [22]. Bcr-Abl causes hyperactivity of Ras, Raf and JAK/STAT, which can occur by multiple mechanisms; i.e., by Bcr-Abl activating these pathways directly, or by the induction of autocrine cytokines, which in turn activate these pathways [23]. Bcr-Abl autophosphorylation of tyrosine 177 provides a docking site for the adapter molecule Grb-2. Grb-2, after binding to the Sos protein, stabilizes Ras in its active GTP-bound form. Two other adapter molecules, Shc and Crkl, can also activate Ras [9, 24]. Ras activates Raf, and finally, Raf initiates a signaling cascade through the serine–threonine kinases Mek1/Mek2 and Erk, which ultimately leads to the transcription of genes involved in cell proliferation and survival (Figure 1), such as c-Myc, Cyclin D, Cyclin A, Bcl-2, cytokines, etc [22].

The JAK/STAT pathway has been demonstrated to be constitutively activated In CML. Among all the molecules participating in these pathways, STAT1 and STAT5 have been found to be the two major STATs phosphorilated by Bcr-Abl. STAT5 has pleiotropic physiologic functions, and its main effect in Bcr-Abl-transformed cells appears to be primarily anti-apoptotic, involving transcriptional activation of Bcl-xL [25]. Also, in some experimental systems there is evidence that Bcr-Abl induces an IL-3 and G-CSF autocrine loop in early progenitor cells [12].

3.2. Inhibition of apoptosis

Leukemic Stem Cells acquire the ability for long-term survival primarily by deregulation of apoptosis. In CML, blocking of apoptosis is mediated by Bcr-Abl. Bcr-Abl may block the release of cytochrome C from mitochondria and thus activation of caspases. This effect upstream of caspase activation might be mediated by the Bcl-2 family of proteins [26]. Bcr-Abl has been shown to up-regulate anti-apoptotic protein Bcl-xL in a STAT5-depend manner, as mention above [27]. Another link between Bcr-Abl and the inhibition of apoptosis might be the phosphorylation of the pro-apoptotic protein Bad through PI3k pathway. Bcr-Abl forms multimeric complexes with PI3 kinase, Cbl, and the adapter molecules Crk and Crkl, in which PI3 kinase is activated. The next substrate in this cascade appears to be the serine-threonine kinase Akt. This kinase had previously been implicated in antiapoptotic signaling and protein Bad as a key substrate of Akt (Figure 1). Phosphorylated Bad is inactive because it is no longer able to bind anti-apoptotic proteins such as Bcl-xL and it is trapped by cytoplasmic 14-3-3 proteins [28].

3.3. Altered adhesion properties

In CML, progenitor cells exhibit decreased adhesion to bone marrow stroma cells and extracellular matrix. From this point of view, adhesion to stroma negatively regulates cell proliferation, and CML cells escape this regulation by virtue of their perturbed adhesion properties. Bcr-Abl directly phosphorylates Crkl, a protein involved in the regulation of cell motility and in integrin-mediated cell adhesion by association with other focal adhesion proteins such as paxillin, the focal adhesion kinase Fak, p130 Cas and Hef1 [29, 30] (Figure 1). In addition to this, it has been demonstrated that the activity of Bcr-Abl promotes expression of integrin β1, a variant not found in the normal counterpart that inhibits adhesion to stroma and cell matrix, together with the effect of expansion and premature exit of myeloid progenitors and precursors to bloodstream [31].

3.4. Self-renewal

Deregulation of self-renewal has been recognized as an important event in disease progression. In normal hematopoietic stem cells, self-renewal capacity involves several signaling pathways: Notch, Wnt, Sonic Hedgehog (Shh), FoxO and Alox5 [32-34].

Notch pathway

Notch receptors are an evolutionarily conserved family of trans-membrane receptors that are known to be expressed and activated in normal HSC. Binding to their physiological ligands, which are part of the Delta and Serrata families, leads to separation of an intracellular portion of Notch. This fragment is capable of entering the nucleus where it binds transcriptional repressor CBF-1. Interconnection of Notch, CBF-1 and the co-factor MAML-1 (mastermind-like-1) leads to transcriptional activation of target genes [35]. Constitutively active Notch is able to mediate multilineage potential *in vivo*. Differentiation of cells leads conversely to downregulation of Notch [36].

Notch signaling may also be important in advanced stages of CML. Hes1, a key Notch target gene, was found to be highly expressed in 8 out of 20 patients with CML in blast crisis, but was not seen in the chronic phase. In mice, the combination of Hes1 and BCR-ABL expression in myeloid lineage progenitor cells resulted in an acute leukemia resembling blast crisis CML [37]. This suggests that Notch inhibitors may be useful in strategies aimed at eradicating CML LSC.

Wnt pathway

In normal hematopoiesis, Wnt pathway activity is required in the bone marrow niche to regulate HSC proliferation and to preserve self-renewal capacity [38]. Activation of the canonical Wnt/β-catenin pathway consists of binding of Wnt proteins to members of the Frizzled and low-density lipoprotein receptor related (LPR) families on the cell surface. In the absence of Wnt signals, β-catenin is associated with a large multiprotein complex that includes Axin, APC, and glycogen synthase kinase 3β (GSK3β), among others. Through a mechanism not entirely understood, when Wnt proteins bind to their target, Axin facilitates phosphorylation of β-catenin by GSK3β. Phosphorylation, in turn, results in ubiquitination, targeting β-catenin for degradation. Thus, axin serves as an inhibitor of β-catenin activity. Binding of Wnt proteins to their receptors leads to activation of Dishevled (Dsh), which inhibits phosphorylation of β-catenin by GSKβ, so it accumulates in the cytoplasm and translocates to the nucleus, where it activates transcription factors, such as LEF/TEF and allows expression of target genes [39].

This pathway has been implicated in CML. Indeed, in blast crisis CML, the LSC, which resemble granulocyte-macrophage progenitor cells (GMP), have aberrant activation of β-catenin via the canonical Wnt signaling pathway. In a proportion of these cases, the pathway is activated through abnormal missplicing of GSK3β [40].

Sonic Hedgehog (Shh) pathway

The Hedgehog (Hh) pathway is a highly conserved developmental pathway, which regulates the proliferation, migration and differentiation of cells during development [41]. It is typically active during development, but silenced in adult tissues, except during tissue regeneration and injury repair [42]. Three distinct ligands, i.e., Sonic (Shh), Indian (Ihh) and Desert (Dhh) Hedgehog exist in humans. Upon ligand binding to the receptor patched (Ptch), inhibition of smoothened (Smo) receptor is relieved. Smo then activates members of the Gli family of zinc-finger transcription factors, which translocate to the nucleus to regulate the transcription of Hh target genes, including Gli1, Gli2, Ptch and regulators of cell proliferation and survival [43].

Based on murine embryonic stem cell studies, it has been found that Hh signaling plays major roles during primitive hematopoiesis. Ihh is a primitive endoderm-secreted signal and is sufficient to activate embryonic hematopoiesis and vasculogenesis [44]. Furthermore, a study of zebrafish showed that the mutations of the Hh pathway members or inhibition of the Hh pathway with the Hh inhibitor cyclopamine can cause a developmental defect in adult HSC [45]. In addition, activation of Hh pathway has been observed in different human cancers. In CML patients, more than four-fold induction of the transcript levels of Gli1 and Ptch was observed in $CD34^+$ cells in both chronic phase and blast crisis. In two studies using a CML mouse model, recipients of the Bcr-Abl transduced bone marrow cells from $Smo^{-/-}$ donor mice developed CML significantly slower than recipients of Bcr-Abl transduced bone marrow cells from wild-type donor mice. When the frequency and function of the LSCs were examined, Smo deletion caused a significant reduction of the percentage or LSCs [46]. By contrast, over expression of Smo led to an increased percentage of LSC and accelerated the progression of CML [47].

FoxO pathway

The FoxO (Forkhead-O) subfamily of transcription factors regulate cell cycle, stress resistance, differentiation, and long-term regenerative potential of HSC [48], and protect integrity of the stem cell pool. There are four members (FoxO1, FoxO3, FoxO4 and FoxO6) and are known to be effectors of the PI3k/AKT pathway, which is frequently mutated or hyperactivated in hematologic malignancies, and are abundantly expressed in the hematopoietic system. Akt directly phosporylates the FoxO members from the nucleus and promotes its degradation in the cytoplasm. FoxO members localize to the nucleus and regulate apoptosis, cell cycle progression and oxidative stress responses [49]. In a model of deficient FoxO mice it was shown a defect in the long-term expansion capacity of the HSC pool. Such a defect has been correlated with increased cell division and apoptosis of HSCs.

FoxO transcription factors have also been shown to have essential roles in the maintenance of CML LSCs [50]. FoxO3 localizes to the cell nucleus and it causes a decrease in Akt phosphorylation in the LSC population. In addition, serial CML transplantation showed that FoxO3 deficiency severely impairs the ability of LSCs to induce CML. Furthermore, transforming growth factor-β (TGF-β) is a crucial regulator of Akt activation and controls FoxO3 localization in LSCs of CML. A combination strategy of TGF-β inhibition, FoxO3 deficiency and Bcr-Abl kinase inhibition results in efficient LSCs depletion and suppression of CML development [51].

Alox5 pathway

The Alox5 pathway is the only one signaling pathway not shared by LSC with normal HSC. The *Alox5* gene encoding arachidonate 5-lipoxygenase (5-LO) is involved in numerous physiological and pathological processes, including oxidative stress response, inflammation and cancer [52]. 5-LO is responsible for producing leukotrienes, a group of inflammatory substances that cause human asthma [53]. Altered arachidonate metabolism by leukocytes and platelets was reported in association with myeloproliferative disorders [54]. Several selective 5-LO inhibitors were found to reduce proliferation and induce apoptosis of CML cells in vitro [55]. Recently, human CML microarray studies have shown that Alox5 is differentially expressed in $CD34^+$ CML cells suggesting a role for Alox5 in human CML stem cells. However, the function of Alox5 in LSCs needs to be tested. Other microarray analysis of gene expression in LSCs in CML mice showed that the ALox5 gene was up-regulated by Bcr-Abl and that this up-regulation was not inhibited by Imatinib treatment, providing a possible explanation of why LSCs are not sensitive to inhibition by Bcr-Abl kinase inhibitors [56].

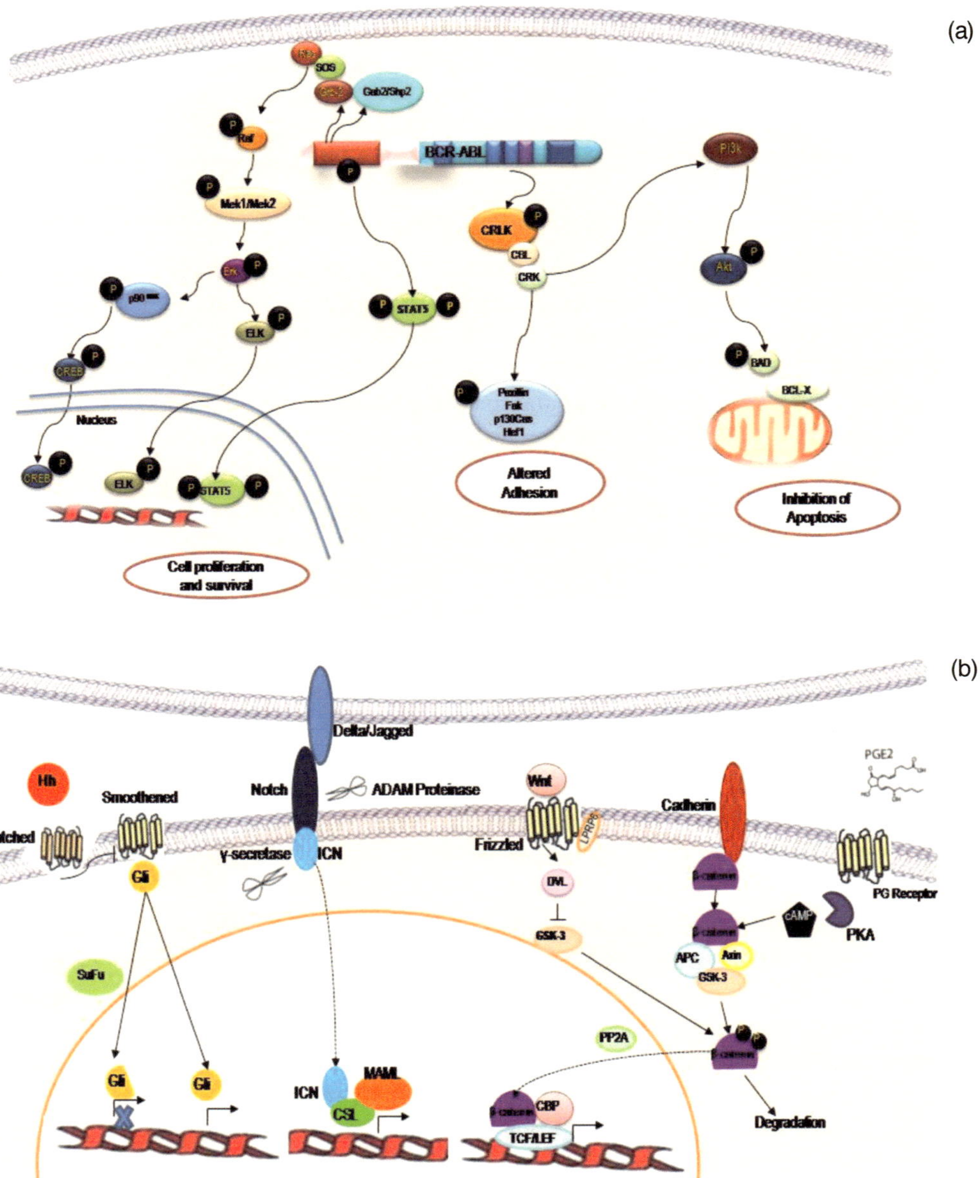

Figure 2. Signaling pathways involved in the signaling of BCR-ABL. A) Schematic representation of principal molecules that participate in proliferation, adhesion and apoptosis. B) Pathways involved in self-renewal.

4. Current therapies

The first effective treatment for CML was the solution of Fowler's, which contained arsenic as active component and was used in the early 20th century. Later between 1920 y1930 irradiation to the spleen was the main therapeutic option, since it offered patients the decrease of symptoms, although it did not prolong their lives. In 1953, busulfan was included in CML treatment. This compound provided benefit in terms of survival, although it was shown to be extremely toxic for hematopoietic progenitor cells. The next drugs effective in the treatment of CML were hydroxyurea and cytosine arabinoside, both less toxic than busulfan and able to block proliferation of cells, but unable to induce specific damage to leukemic cells; thus, patients usually progressed to the accelerated and blast crisis phases [57].

4.1. Interferon-α

Interferon-α (IFNα) was the first drug capable of extending the chronic phase of the disease and retarding the evolution to the accelerated phase. IFNα is a nonspecific stimulant of the immune system that regulates T-cell activity and produces a complete hematologic response (CHR) in 40-80% of patients, and a complete cytogenetic response (CCR) in 6-10% of patients with a median survival of 89 months [58].

In vitro studies have indicated that IFNα might function via selective toxicity against the leukemic clone, since it is able to inhibit long-term cultures from patients with CML in chronic phase and reduces the percentage of Ph+ cells [59]. It also inhibits CML myeloid progenitors while sparing normal myeloid progenitors [60]. In vivo, IFNα enhances immune regulation through the activation of dendritic, natural killer, and cytotoxic T cells, all of them capable of generating anti-tumor responses. In Bcr-Abl+ cells, IFNα induces a state of tumor dormancy and delays progression to advanced phase [61], and is able to modulate hematopoiesis through enhanced adhesion of CML progenitor cells to stromal cells, whereas adhesion of normal progenitors was unaffected. This enhanced adhesion by CML progenitor cells has been associated with a reduction in neuraminic acid levels and by enhanced hematopoietic cell-microenvironmental cell interactions, which is achieved by the induction of molecules such as β2-Integrin, L-selectin, ICAM-1 and ICAM352 [58, 62].

Because IFNα is a nonspecific immunostimulant, it produces secondary symptoms and toxicities and many patients discontinue therapy. However there are evidence that a significant proportion of IFNα-treated patients in prolonged CCR were able to discontinue treatment without disease relapse [63], and it was recently reported that in a specific group of patients treated with monotherapy there are increased numbers of NK cells and clonal $\gamma\delta$ T cells [64].

4.2. Tyrosine kinase inhibitors

Having identified that tyrosine kinase activity of Bcr-Abl is a major factor in the pathophysiology of CML, it was clear that such a molecule was an attractive target for designing a selective kinase inhibitor. In 1996, Buchdunger et al, synthesized several compounds that inhibit the activity of platelet-derived growth factor receptor (PDGF-R) and ABL kinase. One of these was the 2- phenylaminopyrimidine, which served as a starting point for the development of other related compounds [65]. The activity of the 2-phenylaminopyrimidine series was optimized and gave rise to STI571 (also named imatinib mesylate, CGP57148B or Gleevec®, Novartis Pharmaceuticals).

Imatinib

Imatinib is a highly selective inhibitor of the protein tyrosine kinase family, which includes BCR-ABL protein, PDGF-R and the c-kit receptor. It competitively binds to the ATP-binding site of BCR.ABL and inhibits protein tyrosine phosphorylation *in vitro* and *in vivo* [66]. In vitro studies had shown that Imatinib is capable to inhibit cell proliferation of cell lines expressing Bcr-Abl [67-69], effect accomplished through JAK5-STAT and PI3 kinase signaling inhibition [70, 71]. It has also been shown that STI571 can inhibit CML MNC, obtained both in chronic phase and blast crisis [71] and reduces the colony forming cells from Mobilized Peripheral Blood (MPB) and Bone Marrow from patients with CML in chronic phase [60]. Furthermore, Imatinib inhibits proliferation and cell cycle of stem (CD34+CD38-) and progenitor ($CD34^{+}CD38^{+}$) cells without altering the behavior of normal cells [72].

Studies in CML marrow by Holyoake and her colleagues have demonstrated the presence of a rare, highly quiescent, $CD34^{+}$ cell subpopulation in which most of the cells are Ph+ with the ability to proliferate upon specific induction [11]. These cells are insensitive to the effects of STI571 and remain quiescent and viable even in the presence of growth factors [16]. This tumor resistance feature was also reported by Bathia, who mention that STI571 suppressed but does not eliminate primitive cells even after patients remain in CCR [73]. These primitive Ph+ cells could not be detected by nested PCR, when they are obtained from Imatinib-treated patients; however, when the cells are cultured in liquid cultures for a couple of weeks, the Ph+ population becomes detectable, indicating that they were able to remain even after Imatinib treatment [74].

In clinical trials, Imatinib has been shown remarkably effective as a single agent in IFNα-resistant CML chronic phase patients. It induces complete cytogenetic responses in more than 80% of newly diagnosed patients; however, the persistence of detectable leukemic cells in a quiescent state and the presence of patients with resistance or intolerance to Imatinib, lead to the development of a second generation of Tyrosine Kinase Inhibithors.

Nilotinib

Nilotinib (Tasigna, Novartis Pharmaceutical), is an oral aminopyrimidine that is a structural derivative of Imatinib. It was designed to be more selective against the Bcr-Abl tyrosine kinase than imatinib. Like imatinib, it acts through competitive inhibition of the ATP site in the kinase domain [75]. Clinically Nilotinib showed activity in imatinib-resistant patients in all phases of the disease. In chronic phase, it induced 92% of CHR and in accelerated phase and blast crisis the hematological responses were achieved in 72% of cases [76].

In vitro, Nilotinib is 20 times more potent than imatinib against cells expressing wild type Bcr-Abl, and similar results have been observed in studies of mutants cell lines, with the exception of the T315I mutation, which is resistant to both TKIs [77]. In primary CML CD34+ cells,

Imatinib-induced apoptosis is preceded by Bim accumulation; this effect was decreased when cells were cultured in a cytokine-containing medium [78]. In contrast to Imatinib, whose main effect on CML cells seems to be induction of apoptosis, the predominant effect of nilotinib seems to be antiproliferative -rather than apoptotic [17]. Indeed, it has been suggested that Nilotinib can induce a G_0/G_1 cell cycle blockade in cells expressing wild type Bcr-Abl, which could result in disease persistence [79].

Dasatinib

Dasatinib (Sprycel, Bristol-Myers Squibb) is a potent, orally bioavailable thiazolecarboxamide. It is structurally unrelated to imatinib; it has the ability to bind to multiple conformations of the Abl kinase domain and it also inhibits SRC family kinases. In vitro, Dasatinib demonstrated 325-fold greater activity against native Bcr-Abl, as compared with imatinib, and it has shown efficacy against all imatinib-resistant Bcr-Abl mutants with the exception of T351I. Dasatinib is also active against PDGFR, C-Kit and ephrin A receptor [75, 76].

Dasatinib is very effective at inducing apoptosis in CML cells –either, in the presence or absence of added growth factors- and in contrast to Imatinib, that kills those cells destined to move from G_0/G_1 cell cycle phases, but is unable to act on those cells destined to remain quiescent in culture, Dasatinib can act on quiescent $CD34^+$ cells. As expected, based on its structure and mode of action, it has selective cytotoxic activity for leukemic cells over normal cells [80].

Other tyrosine kinase inhibitors

Several TKIs have been developed that exhibit a target spectrum similar to the approved drugs, although they are distinct in terms of off-target effects [81].

SKI-606 (Bosutinib)

Bosutinib (Wyeth) is a 4 anilino-3-quinolinecarbonitrile dual inhibitor of Src and Abl kinases without effect in c-Kit or PDGFR. It has 200-fold grater potency for Bcr-Abl than imatinib and has activity against a number of mutations, but not T315I [76]. In clinical trials, Bosutinib induced 73% of complete hematological response in patients pretreated with Imatinib followed by Dasatinib [82]. In vitro, Bosutinib effectively inhibits Bcr-Abl kinase activity and Src phosphorylation, and reduces the proliferation and CFC growth in CML CD34+ cells; however, it does not seem to induce apoptosis [19].

AP24534 (Ponatinib)

Ponatinib, is a mulitargeted kinase inhibitor that is active against all BCR-ABL mutants, including T315I. This drug also inhibits FLT3, FGFR, VEGFR, c-Kit, and PDGFR and is able to reduce the proliferation of different cell lines and prolong survival of mice that have been injected intravenously with BCR-ABL. Ponatinib showed significant activity in a phase I study of patients with Ph+ cells who had failed to other TKIs [81, 83].

4.3. Hematopoietic cell transplant

Although molecular therapy for CML is highly effective and generally non-toxic, it is unclear whether long-term outcomes with the different therapies (IFNα or TKIs) will be equivalent to cases treated with allogeneic stem cell transplantation, which has shown the highest percentage of long-term disease-free survival of any therapy [75].

In patients younger than 50 years of age and who receive a transplant before 1 year after diagnosis, 5 years survival rates superior to 70% have been attained. However, the application of this procedure is limited by the availability of matched donors and by the toxicity of the procedure in older patients. Moreover, outcomes deteriorate with disease duration [76]. This information associated with the knowledge that quiescent leukemic stem cells remain in patients after treatment, several other agents has been reported.

4.4. Other agents

Danusertib (PHA 739358) is a small molecule with activity against BCR-ABL and aurora kinases and it is able to block the proliferation of leukemia cell lines as well as CD34+ cells from newly diagnosed CML patients including the mutation T315I. However, similarly to other tyrosine kinase inhibitors, no induction of apoptosis in quiescent hematopoietic stem cells could be achieved and resistant BCR-ABL positive clones emerged in the course of Danusertib treatment. This latter observation is related to Abcg2 proteins over-expression [84].

Lonafarnib (SCH66336) is an orally bioavailable non peptidomimetic farnesyl trransferase inhibitor with significant activity against Bcr-Abl+ cell lines and primary CML cells. It can enhance the toxicity of Imatinib in K562 cell line and can inhibit the proliferation of imatinib-resistant cells and increases imatinib-induced apoptosis. However it is unable to kill quiescent CD34+ leukemic cells [20]. In a clinical phase 1 study, it was shown that the combination of Lonafarnib and Imatinib is well tolerated in patients with CML who failed Imatinib, with some patients achieving a complete hematologic response and a complete cytogenetic response [85].

INNO 406 is a 2 phenylaminopyrimidine Bcr-Abl inhibitor with activity against PDGF, c-kit and Lyn that have shown to be 25-55 times more potent than Imatinib in Bcr-Abl+ cell lines. In contrast to other molecules INNO406 does not inhibit all SRC kinases, but it induces programmed cell death in chronic myelogenous leukemia (CML) cell lines through both caspase-mediated and caspase-independent pathways [86].

MK0457 is an aurora kinase inhibitor with activity against Bcr-Abl. This agent was observed to inhibit autophosphorylation of T315I mutant and demonstrate antiproliferative effects in CML cells derived from patients with this mutation, an event that may lead to its use as a combination partner with the approved and established TKI [76].

5. TKI resistance mechanisms

The knowledge of the central role of BCR-ABL in the pathogenesis of CML has allowed the development of several drugs that inhibit the constitutive activity of such an ABL tyrosine kinase. However, although the treatment with tyrosine kinase inhibitors has proven effective in about 80% of CML patients at any stage, the remaining 20% can't respond to it [87].

In CML, the criteria for successful response to treatment, as established by the European consortium LeukemiaNet and subsequently adopted by the National Comprehensive Cancer Network (NCCN) [88], include: complete hematologic remission (CHR), that is to say, a normal blood cell count and complete disappearance of signs and symptoms of the disease; complete cytogenetic response (CCR), which means the total absence of Ph+ metaphases; and complete molecular response, in which transcripts for BCR-ABL are no longer detectable. Using these response criteria, drug resistance is defined as the inability to achieve any of the following: a complete hematologic response (CHR) at 3 months, any cytogenetic response (CyR) at 6 months, partial cytogenetic response (PCyR) at 12 months, or a complete cytogenetic response (CCR) at 18 months of treatment with Imatinib [89].

Two types of resistance mechanisms to TKIs have been described: 1) Primary resistance, which occurs in less than 10% of cases and is defined as the failure of therapeutic effect during the chronic phase of CML without changing clones; and 2) secondary resistance, defined as the loss of the response initially obtained, and commonly occurs in accelerated phase (40-50%) and blast (80%) [90]

It is estimated that the probability of an individual to stay in CCR for 5 years after diagnosis, after treatment with Imatinib is approximately 63%; however, this percentage may represent a sub-estimation since in a significant proportion of cases there is discontinuation of treatment and this, of course, may underestimate the efficacy of the drug [91].

The molecular mechanisms of acquired drug resistance can be divided into two categories: BCR-ABL-dependent and BCR-ABL-independent.

5.1. Bcr-Abl-dependent resistance mechanisms

The inhibition of the activity of tyrosine kinase turned out to be an ideal target for molecular therapy in CML [67]. However, shortly after the introduction of Imatinib, in vitro studies demonstrated that some cell lines became refractory to the drug, suggesting a possible inherent or acquired resistance to therapy [92]. This was quickly followed by the clinical description of patients resistant to Imatinib.

BCR-ABL mutations

The most common mechanism against TKIs therapy are point mutations within the kinase domain, which make conformational changes that decrease the affinity of the TKIs to BCR-ABL kinase domain. These point mutations in the *BCR-ABL* kinase domain are a major cause of Imatinib resistance, and may be identified in approximately 50% or more of the cases. Many more than 100 different mutations affecting more than 70 amino acids have so far been identified, with varying degrees of clinical relevance [93].

The first point mutation reported in TKI resistance was in the region coding for the ATP-binding site of the ABL kinase domain resulting in a threonine to isoleucine substitution at amino acid 315 (Th315→Ile315; T315I) preventing the formation of a hydrogen bond between the oxygen atom provided by the side chain of threonine 315 and the secondary amino group of Imatinib. Moreover, isoleucine contains an extra hydrocarbon group on its side chain, and this inhibits the binding of Imatinib [94]. T315I confers resistance to all currently approved BCR–ABL kinase inhibitors. Recent reports have shown that T315I mutation can be found in approximately 15% of patients after failure of imatinib therapy [85].

Other important TKI's resistant mutations are frequently mapped to the P-loop region (residues 244 to 256) of the kinase domain, which serves as a docking site for phosphate moieties of ATP and interacts with imatinib through hydrogen and van der Waals bonds. These mutations modify the flexibility of the P-loop and destabilize the conformation required for Imatibib binding [95]. Clinical relevance of P-loop mutations is that imatinib treated patients who harbor them have been suggested to have a worse prognosis than those with non-P-loop mutations [96]. Another study identified BCR/ABL mutations in $CD34^+$ cells from CML patients in CCR following Imatinib treatment and suggested that these mutations could lead to imatinib resistance in a small population of progenitors, which consequently could expand and cause the relapse [97].

Several additional mutations that disrupt the interaction between TKIs and BCR-ABL have been characterized, including the P-loop, C-helix, SH2 domain, substrate binding site, A-loop, and C-terminal lobe, some even prior to the initiation of therapy [98]. Most of the reported mutants are rare, however seven mutated sites constitute two thirds of all detected mutations: G250, Y253, E255 (P loop), T315I (gatekeeper), M351, F359, and H396 (activation loop or activation loop backbone) and are frequently evident in the later disease stages [99]. Recently a pan-BCR-ABL inhibitor active against the native enzyme and all tested resistant mutants, including the uniformly resistant T315I mutation has been developed [100].

BCR-ABL kinase domain mutations are not induced by the drug, but rather, just like antibiotic-resistance in bacteria, arise through a process whereby rare pre-existing mutant clones are self-selected due to their capacity to survive and expand in the presence of the drug thus gradually outgrowing drug-sensitive cells [101].

BCR-ABL gene amplification

Overexpression of Bcr-Abl leads to resistance by increasing the amount of target protein needed to be inhibited by the therapeutic dose of the drug. Amplification of the BCR–ABL gene was first described in resistant CML cell lines generated by serial passage of the cells in Imatinib containing media and demonstrated elevated Abl kinase activity due to a genetic amplification of the Bcr–Abl sequence [102, 103].

Cells expressing high amounts of Bcr-Abl in $CD34^+$ CML cells, as in blast crisis, are much less sensitive to Imatinib and, more significantly, take a substantially shorter time for yielding a mutant subclone resistant to the inhibitor than cells with low expression levels, as in chronic phase [104]. However overexpression and amplification of the *BCR-ABL* gene itself accounts for Imatinib failure in a smaller percentage of patients with an overall percentage of 18% [94].

http://dx.doi.org/10.5772/54651

5.2. BCR-ABL-independent resistance mechanisms

Drug efflux

HSC are characterized by their ability to pump-out fluorescent dyes, and this led to isolation of stem cells based on this property. In fact, such an efflux capacity has become one of the most efficient methods to purify stem cells from different sources [105]. In this regard, ATP-binding cassette (ABC) transmembrane transporters have shown to be responsible for most of the efflux of the fluorescent dyes in HSCs [106].

In cancer cell lines, multidrug resistance is often associated with an ATP-dependent decrease in cellular drug accumulation, which is attributed to the overexpression of ABC transporter proteins [107]. The first studies on imatinib-resistance showed increased levels of the multidrug resistance protein MDR1 (ABCB1) in Imatinib resistant BCR-ABL+ cell lines [108]. Later on, it was confirmed that Imatinib is a substrate of membrane ABC transporters, such as ABCB1 (MDR1, P-gp), and that variations in the activity or expression of P-gp affects the pharmacokinetics of Imatinib, reducing or increasing its bioavailability [109]. P-gp-positive leukemic cells have low intracellular levels of Imatinib; decreased Imatinib levels, in turn, were associated with a retained phosphorylation pattern of the Bcr-Abl target Crkl and loss of effect of Imatinib on cellular proliferation and apoptosis. The modulation of P-gp by Ciclosporin A readily restored imatinib cytotoxicity in these cells [110].

Another drug efflux pump, the breast cancer resistance protein BRCP encoded by ABCG2, has also been implicated in Imatinib resistance. Imatinib has been variably reported to be a substrate and/or an inhibitor for the BCRP/ABCG2 drug efflux pump, which is overexpressed in many human tumors and also found to be functionally expressed in CML stem cells [111, 112].

CML stem cells have been shown to express the ATP dependent transporter cassette protein ABCG2, which could decrease the intracellular accumulation of Imatinib in CML LSC [103]. Thus, overexpression of ABC transporters gives protection to tumor cells from TKIs [114].

Drug intake

Inversely to the drug efflux pump proteins, the human organic cation transporter 1 (OCT1) mediates the active transport of Imatinib into cells, and inhibition of OCT1 decreases the intracellular concentration of Imatinib [115]. OCT1 was also found to be expressed in significantly higher levels in patients who achieved a CCR to Imatinib than in those who were more than 65% Ph chromosome positive after 10 months of treatment [116]. Tyrosine Kinase Inhibitor Optimization and Selectivity (TOPS) trial suggested that patients with lower hOCT1 levels had reduced MMR rates at 12 months when receiving the standard dose of Imatinib, compared with high-dose Imatinib [117].

Recently Engler and cols. found that the intracellular uptake and retention (IUR) of imatinib, OCT-1 activity and OCT-1 mRNA expression are all significantly lower in CML $CD34^{+}$ cells. However, no differences in IUR or OCT-1 activity were observed between these subsets in healthy donors. Low Imatinib accumulation in primitive CML cells, mediated through reduced OCT-1 activity may be a critical determinant of long-term disease persistence [118].

Differential interactions between drug efflux/influx pumps and kinase inhibitors might be a possible means to tailor drug selection for individual patients, because OCT-1 expression is a key determinant of intracellular availability of Imatinib but not of Nilotinib [119]. Other TKIs, such as Dasatinib and, as just mentioned, Nilotinib, do not appear to be substrates for hOCT1, but whether this difference alone will lead to reduced resistance rates with these second-generation TKIs remains unknown [120]. An adequate balance between influx (hOCT1) and efflux (MDR1, ABCG2) transporters may be a critical determinant of intracellular drug levels and, hence, resistance to Imatinib.

Quiescence

One feature of CML is the presence of a population of highly quiescent primitive cells [11], which, as their normal counterparts, is capable of regenerating hematopoiesis and reconstitutes the disease in immunocompromised mice [121]. These stem cells are Ph+, express high levels of CD34 and do not express CD38, CD45RA and CD71, and may spontaneously exit the G_0 phase and enter a state of constant proliferation [122]. Several reports have documented that quiescent cells from CML patients are insensitive to in vitro treatment with Imatinib and Dasatinib [16, 123].

A possible cause of insensitivity to TKIs is that BCR-ABL mRNA transcript levels are 300-fold higher in the most primitive $CD34^+CD38^-Lin^-$ population than in terminally differentiating $CD34^-Lin^+$ CML cells [124]. It has been reported that elevated levels of Bcr-Abl confer reduced sensitivity to Imatinib [125]. Moreover, the quiescent state of CML stem cells allows them to evade chemotherapy treatments, which are designed to eliminate metabolically active cell population as well as targeted therapies, thus contributing to relapse when treatment with tyrosine kinase inhibitors is discontinued.

Activation of BCR-ABL alternative signaling

BCR-ABL activates different signaling pathways that promote the growth and survival of hematopoietic cells, thus inducing cell transformation. These pathways include Ras, mitogen activated protein kinase (MAPK), c-jun N-terminal kinase (JNK), stress-activated protein kinase (SAPK), nuclear factor kappa B(NF-kB), signal transducers and activators of transcription (STAT), phosphoinositide 3- (PI-3) kinase, and c-Myc [126]. A well characterized pathway involves the Src Family Kinases (SFKs), which are activated by BCR-ABL and the subsequent inhibition of BCR-ABL by Imatinib may not result in the complete inhibition of Src family kinases elucidating a Bcr-Abl independent mechanism of imatinib resistance [127]. Phosphorylation of the Bcr-Abl SH2 and SH3 domains by the SFK may increase the activity of the Abl kinase and may alter its susceptibility to Imatinib [128].

Activation of the Janus kinase (Jak) and subsequent phosphorylation of several Signal Transducer and Activator of Transcription (STAT) family members has been identified in both Bcr–Abl–positive cell lines and in primary CML cells and may contribute to the transforming ability of Bcr–Abl [129].

The tyrosine residue at position 177 within the BCR portion is essential for the binding of adaptor proteins, including Growth Factor Receptor-Bound Protein 2 (GRB2) GRB10, 14-3-3, and the SH2 domain of ABL1 [130]. Bcr-Abl protein is able to activate the Ras/Raf/Mek kinase pathway and the phosphatidylinositol 3′ kinase (PI3K)/Erk pathways through GRB2 [131, 132].

Autocrine loops could contribute to resistance. It has been demonstrated that IL-3 and granulocyte-colony G-CSF are produced within primitive CD34+ cells from patients with CML-CP, both of these cytokines stimulate cellular proliferation in an autocrine manner and protect cells from Imatinib-induced apoptosis [122].

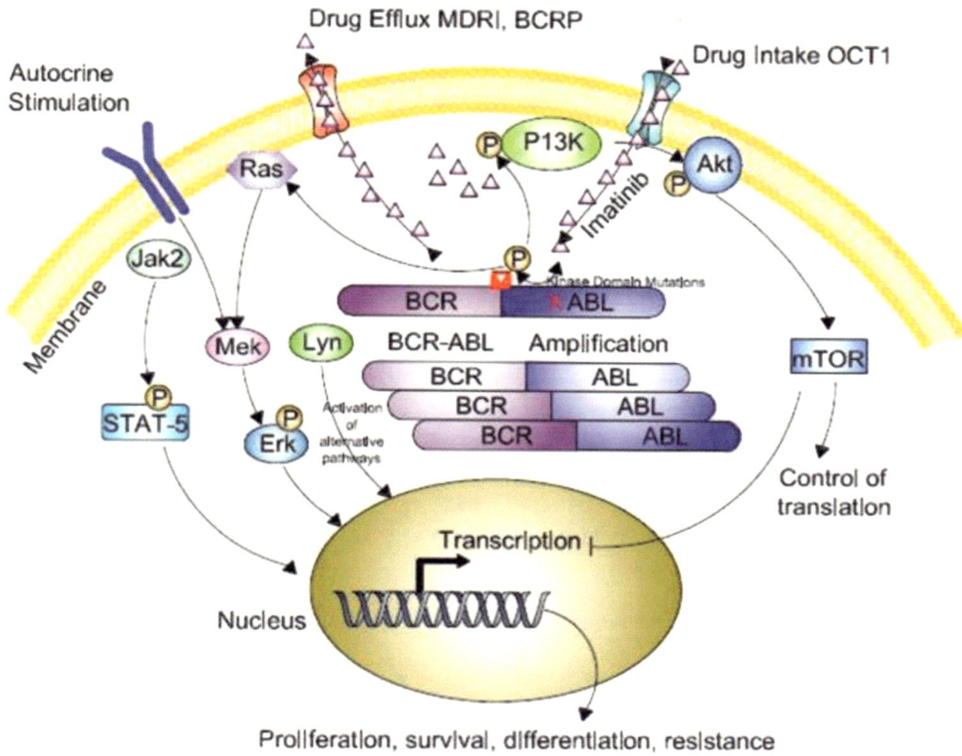

Figure 3. Resistance mechanism in Chronic Myeloid Leukemia. Principal mechanisms involved in dependent and independent BCR-ABL mechanisms are shown (modified to [99]).

6. Concluding remarks

The presence of a rare population of cells capable of initiating and sustaining leukemia in CML (LSC) has major implications for the biology of the disease and the development of new and more effective treatments. As recognized by several investigators, LSC are key players in the origin and progression of CML, as well as in the reappearance of the disease after treatment. Thus, it is evident that novel therapies must be directed towards the elimination of such cells. However, since their numbers within the marrow microenvironment are extremely low, as compared to the bulk of the malignant cells, and their biology is quite different from that of the rest of the CML cells, the task of finding solutions to this problem is a rather difficult one. It is a great challenge, but significant advances will surely be achieved in the years to come.

Acknowledgements

Antonieta Chàvez-González is recipient of funding from the National Council of Science and Technology CONACYT (grant CB 2008-01-105994) and from the Mexican Institute for Social Security IMSS (grant IMSS/PROT/G11/946). Dafne Moreno-Lorenzana and Socrates Avilés-Vazquez are scholarship holders from CONACYT and IMSS. Héctor Mayani is a scholar of FUNDACION IMSS (Mexico) and his research is supported by grants from the National Council of Science and Technology-CONACYT (grant SALUD-69664).

Author details

Antonieta Chávez-González*, Sócrates Avilés-Vázquez, Dafne Moreno-Lorenzana and Héctor Mayani

*Address all correspondence to: achavez_g@yahoo.com.mx

Oncology Research Unit, Oncology Hospital, Mexican Institute for Social Security, Mexico City, Mexico

References

[1] Sawyers, C. L. Chronic myeloid leukemia. New England Journal of Medicine (1999). , 340, 1330-1340.

[2] Rowley, J. A new consistent chromosomal abnormality in chronic myelogenous leukaemia identified by quinacrina fluorescence and Giemsa staining. Nature (1973). , 243, 290-293.

[3] Piller, G. Leukaemia- A brief historical review from ancient times to 1950. British Journal of Haematology (2001). , 112, 282-292.

[4] Tsao, A, Kantarjian, H, Talpaz, M, et al. STI571 in Chronic Myelogenous Leukaemia. British Journal of Haematology (2002). , 119-24.

[5] Crews, L, & Jamieson, C. Chronic Myeloid Leukemia Stem Cell Biology. Current Hematologic Malignant Reports (2012). , 7, 125-132.

[6] Chávez-gonzález, A, Ayala-sanchez, M, Mayani, H, et al. La leucemia Mieloide Crónica en el Siglo XXI. Biologìa y Tratamiento. Revista de Investigación Clínica (2009). , 61, 221-232.

[7] Faderl, S, Talpaz, M, Estrov, Z, et al. The Biology of Chronic Myeloid Leukemia. The New England Journal Medicine (1999). , 341, 164-172.

[8] Albano, F, Anelli, L, Zagaria, A, et al. Non random distribution of genomic features in breakpoint regions involved in chronic myeloid leukemia cases with variant t(9;22) or additional chromosomal rearrangements. Molecular Cancer. (2010). , 9, 120-135.

[9] Deininger, M, Goldman, J, & Melo, J. The molecular biology of chronic myeloid leukemia. Blood (2000). , 96, 3343-3356.

[10] Quintas-cardama, A, & Cortes, J. Molecular biology of bcr-abl positive chronic myeloid leukemia. Blood (2009). , 1.

[11] Holyoake, T, Jiang, X, Eaves, C, & Eaves, A. Isolation of a highly quiescent subpopulation of primitive leukemic cells in chronic myeloid leukemia. Blood (1999). , 2056-2064.

[12] Jiang, X, Lopez, A, Holyoake, T, et al. Autocrine production and action of IL-3 and granulocyte colony-stimulating factor in chronic myeloid leukemia. Proceedings of the National Academy of Sciences (1999). , 96, 12804-12809.

[13] Lemoli, R, Salvestrini, V, Bianchi, E, et al. Molecular and functional analysis of the stem cell compartment of chronic myelogenous leukemia reveals the presence of a CD34- cell population with intrinsic resistance to imatinib. Blood (2009). , 114, 5191-5200.

[14] Daley, G, Van Etten, R, Baltimore, D, et al. Induction of chronic myelogenous leukemia in mice by the bcr-abl gene of the Philadelphia chromosome. Science (1990). , 210.

[15] MayaniFlores-Figueroa, Chavez-Gonzalez. In vitro biology of human myeloid leukemia. Leukemia Research. (2009). , 33, 624-637.

[16] Graham, S, Jorgenssen, H, Allan, E, et al. Primitive quiescent Philadelphia positive stem cells from patients with chronic myeloid leukemia are insensitive to STI571 in vitro. Blood (2002). , 99, 319-325.

[17] Jorgensen, H, Allan, E, et al. Nilotinib exerts equipotent antiproliferative effects to imatinib and does not induce apoptosis in CD34 CML cells. Blood (2007). , 109, 4016-4019.

[18] Copland, M, Hamilton, A, Erlick, L, et al. Dasatinib (BMS354825) targets an earlier progenitor population than imatinib in primary CML but does not eliminate the quiescent fraction. Blood (2006). , 107, 4532-4539.

[19] Konig, H, Holyoake, T, Bhatia, R, et al. Effective and selective inhibition of chronic myeloid leukemia primitive hematopoietic progenitors by the dual Src/Abl kinase inhibitor SKI 606. Blood (2008). , 111, 2329-2338.

[20] Jorgensen, H, Allan, E, Graham, S, et al. Lonafarnib reduces the resistance of primitive quiescent CML cells to imatinib mesylate in vitro. Leukemia (2005). , 19, 1184-1191.

[21] Hamilton, A, Helgason, V, Schemionek, M, et al. Chronic myeloid leukemia stem cells are not dependent on Bcr-Abl kinase activity for their survival. Blood (2012). , 119, 1401-1510.

[22] Steelman, L. S, Pohnert, S. C, Shelton, J. G, et al. JAK/STAT, Raf/MEK/ERK, PI3K/Akt and BCR-ABL in cell cycle progression and leukemogenesis. Leukemia (2004). , 18, 189-218.

[23] Liu, R, Fan, C, Garcia, R, et al. Constitutive activation of the JAK2STAT5 signal transduction pathway correlates with growth factor independence of megakaryocytic leukemia cell lines. Blood.(1999). , 93, 2369-2379.

[24] OdaHeaney C, Hagopian JR, et al. Crkl is the major tyrosine-phosphorylated protein in neutrophils of patients with CML. Journal of Biological Chemistry (1994). , 269, 22925-22928.

[25] Ilarioa R JrVan Etten R. and P190 (BCR/ABL) induce the tyrosine phosphorylation of SHc proteins in human tumors. Oncogene (1995). , 210.

[26] Amarante Mendes GNaekygung C, Liu L, et al. Bcr-Abl exerts its antiapoptotic effect against diverse apoptotic stimuli through blockade of mitochondrial release of cytochrome C and activation of caspase-3. Blood (1998). , 91, 1700-1705.

[27] Skorski, T, Bellacosa, A, Nieborowska-skorska, M, et al. Transformation of hematopoietic cells by BCR/ABl requires activation of PI3k/Akt depend pathway. EMBO Journal.(1997). , 16, 6151-6161.

[28] Zha, J, Harada, H, Yang, E, et al. Serine phosphorylation of death agonist BAD in response to survival factor results in binding to 14-3-3 not BCL-xL. Cell (1996). , 87, 619-628.

[29] Salgia, R, Pisick, E, Sattler, M, et al. forms a signaling complex with the adapter protein CRKL in hematopoietic cells transformed by the BCRABL oncogene. Journal of Biological Chemistry(1996). , 130CAS.

[30] Sattler, M, Salgia, R, Shrikhande, G, et al. Differential signaling after β1 integrin ligation is mediated through binding of CRKL to CBL) and p110 (HEF1). Journal of Biological Chemistry (1997). , 210.

[31] Zhao, R, Tarone, G, & Verfaillie, C. Presence of the adhesion inhibitory β1B integrin isoform on CML but not normal progenitors is at least in part responsable for the decrease CML progenitor adhesión. Blood.(1997). a.

[32] Heidel, F, Mar, B, & Armstrong, S. Self-renewal related signaling in myeloid leukemic stem cells. International Journal of Hematology (2011). , 94(2), 109-117.

[33] Ciloni, D, & Saglio, G. Molecular Pathways: BCR-ABL. Clinical Cancer Research (2012). , 18(4), 1610-1613.

[34] Sloma, I, Jiang, X, Eaves, A. C, et al. Insights into the stem cells of chronic myeloid leukemia. Leukemia (2010). , 24, 1823-1833.

[35] Duncan, A. W. Integration of Notch and Wnt signaling in hematopoietic stem cell maintenance. Nature Immunology (2005). , 6(3), 314-322.

[36] Varnum-finney, B, et al. Immobilization of Notch ligand, Delta-1, is required for induction of Notch signaling. Journal of Cell Science (2000). , 133, 4313-4318.

[37] Klinakins, A, et al. A novel tumor-supressor function for the Notch pathway in myeloid leukaemia. Nature (2011). , 473, 230-233.

[38] Reya, T, Duncan, A. W, Ailles, L, et al. A role for Wnt signalling in self-renewal of haematopoietic stem cells. Nature (2003). , 423, 409-414.

[39] Ikeda, S, Kishida, S, Yamamoto, H, et al. Axin, a negative regulator of the Wnt signaling pathway, forms a complex with GSK-3beta and beta-catenin and promotes GSK- 3beta-dependent phosphorylation of beta-catenin. EMBO Journal (1998). , 17, 1371-1384.

[40] Zhao, C, Blum, J, Chen, A, et al. Loss of beta-catenin impairs the renewal of normal and CML stem cells in vivo. Cancer Cell (2007). , 12, 528-541.

[41] Ingham, P. W, & Mcmaon, A. P. Hedgehog signaling in animal development: paradigms and principles. Genes and Development (2001). , 15, 3059-3087.

[42] Ahn, S, & Joyner, A. L. In vivo analysis of quiescent adult neural stem cells responding to Sonic hedgehog. Nature. (2005). , 437, 324-331.

[43] Teglund, S, & Toftgard, R. Hedgehog beyond medulloblastoma and basal cell carcinoma. Biochimica et Biophysica Acta (2010). , 1805, 182-208.

[44] Dyer, M. A, Farrington, S. M, Mohn, D, et al. Indian hedgehog activates hematopoiesis and vasculogenesis and can respecify prospective neurectodermal cell fate in the mouse embryo. Development (Cambridge, England) (2001). , 128, 1717-1730.

[45] Gering, M, & Patient, R. Hedgehog signaling is required for adult blood stem cell formation in zebrafish embryos. Development Cell (2005). , 8, 389-400.

[46] Dierks, C, Beigi, R, Guo, G. R, et al. Expansion of Bcr-Abl-positive leukemic stem cells is dependent on Hedgehog pathway activation. Cancer Cell (2008). , 14, 238-249.

[47] Blum, J, et al. Hedgehog signaling is essential for maintenance of cancer stem cells in myeloid leukemia. Nature. (2009). , 458, 776-779.

[48] Thotova, Z, & Gilliland, D. G. FoxO transcription factors and stem cell homeostasis: insights from the hematopoietic system. Cell Stem Cell (2007). , 1, 140-152.

[49] Tothova, Z, et al. FoxOs are critical mediators of hematopoietic stem cell resistance to physiologic oxidative stress. Cell (2007). , 128, 325-339.

[50] Naka, K, et al. TGF-β-FOXO signaling pathway maintains leukemia-initiating cells in chronic myeloid leukemia. Nature (2010). , 463, 676-680.

[51] Naka, K, Hoshii, T, Muraguchi, T, et al. TGF-beta-FOXO signaling maintains leukaemia-initiating cells in chronic myeloid leukaemia. Nature (2010). , 463, 676-680.

[52] Catalano, A, Rodilossi, S, Caprari, P, et al. Lipoxygenase regulates senescence-like growth arrest by promoting ROS-dependent activation. EMBO Journal (2005). , 53.

[53] Peters-golden, M. Henderson Jr WR. Leukotrienes. New England Journal of Medicine (2007). , 357, 1841-1854.

[54] Takayama, H, Okuma, M, Kanaji, K, et al. Altered arachidonate metabolism by leukocytes and platelets in myeloproliferative disorders. Prostaglandines Leukotrienes and Medicine (1983). , 12, 261-272.

[55] Anderson, K, Seed, T, Plate, J, et al. Selective inhibitors of 5-lipoxygenase reduce CML blast cell proliferation and induce limited differentiation and apoptosis. Leukemia Research (1995). , 19, 789-801.

[56] Chen, Y, Hu, Y, Zhang, H, et al. Loss of the Alox5 gene impairs leukemia stem cells and prevents chronic myeloid leukemia. Nature Genetics (2009). , 41, 783-792.

[57] Fausel, C. Targeted chronic myeloid leukemia therapy. Seeking a cure. American Journal of Health System Pharmacy (2007). SS15., 9.

[58] Dowding, C, Gordon, M, Guo, A, et al. Potential mechanisms of action of interferon-alpha in CML. Leukemia and Lymphoma (1993). , 11, 185-191.

[59] Cornelissen, J, Ploemacher, R, Wognum, B, et al. An in vitro model for cytogenetic conversion in CML. Interferon-alpha preferentially inhibits the outgrowth of malignant stem cells preserved in long-term culture. Journal of Clinical Investigation (1998). , 102, 976-983.

[60] Marley, S, Deininger, M, Davidson, J, et al. The tyrosine kinase inhibitors STI561, like interferon-alpha, preferentially reduces the capacity for amplification of granulocyte-macrophage progenitors from patients with chronic myeloid leukemia. Experimental Hematology (2000). , 28, 551-557.

[61] Talpaz, M. Interferon-alfa-based treatment of chronic myeloid leukemia and implications of signal transduction inhibition. Seminars in Hematology (2001). , 38, 22-27.

[62] Martín-henao, G, Quiroga, R, Sureda, A, et al. L-selectin expression is low on CD34+ cells from patients with chronic myeloid leucemia and interferon-a up regulates this expression. Haematologica (2000). , 85, 139-146.

[63] Bonifazi, F, De Vivo, A, Rosti, G, et al. Chronic myeloid leukemia and interferon alpha: a study of complete cytogenetic responders. Blood (2001). , 98, 3074-3081.

[64] Kreutzman, A, Rohon, P, Faber, E, et al. Chronic Myeloid Leukemia Patients in prolonged remission following interferon a monotherapy have distinct cytokine and oligoclonal lymphocyte profile. Plos One (2011). , 6, 1-12.

[65] Bochdunger, E, Zimmermann, J, Mett, H, et al. Inhibition of the Abl protein-tyrosine kinase in vitro and in-vivo by a 2-phenylamiropyrimidine derivative. Cancer Research (1996). , 56, 100-104.

[66] Schindler, T, Bornmann, W, Pellicena, P, et al. Structural mechanism for STI571 inhibition of Abelson tyrosine kinase. Science (2000). , 289, 1938-1942.

[67] Druker, B, Tamura, S, Buchdunger, E, et al. Efects of a selective inhibitor of the Abl tyrosine kinase on the growth of Bcr-Abl positive cells. Nature Medicine (1996). , 2, 561-566.

[68] Carroll, M, Ohno-jhones, S, & Tamura, S. CGP57148 a tyrosine kinase inhibitor, inhibits the growth of cells expressing BCR-ABL, REL-ABL, and TEL-PDGFR fusion proteins. Blood (1997). , 90, 4947-4952.

[69] Benjamin, M, Chandra, J, Svingen, P, et al. Effects of the Bcr/abl kinase inhibitors STI571 and adaphostin NSC680410 on CML cells in vitro. Blood (2002). , 99, 664-671.

[70] Kindler, T, Breitenbuecher, F, Kasper, S, et al. In BCR-ABL positive cells, STAT5 tyrosine-phosphorylation integrates signal induced by imatinib mesylate and AraC. Leukemia (2003). , 17, 999-1009.

[71] Gambacorti-passerini, C, Barni, R, Marchsi, E, et al. Sensitivity to the abl inhibitor STI571 in fresh leukaemic cells obtained from chronic myelogenous leukemia patients in different stages of disease. British Journal of Haematology (2001). , 112, 972-974.

[72] Holtz, M, Slovak, M, Zhang, F, et al. Imatinib mesylate (STI571) inhibits growth of primitive malignant progenitors in chronic myelogenous leukemia through reversal of abnormally increased proliferation. Blood (2002). , 99, 3792-3800.

[73] Bhatia, R, Holtz, M, Niu, N, et al. Persistence of malignant hematopoietic progenitors in chronic myelogenous leukemia patients in complete cytogenetic remission following imatinib mesylate treatment. Blood (2003). , 101, 4701-4707.

[74] Chavez-gonzalez, A, Ayala-sanchez, M, Sanchez-valle, E, et al. Functional integrity in vitro of hematopoietic progenitor cells from patients with chronic myeloid leukemia that have achieved hematological remission after different therapeutic procedures. Leukemia Research (2006). , 30, 286-295.

[75] Shah, N. Loss of response to imatinib: Mechanisms and management. Hematology (2005). , 183-187.

[76] Ramirez, P. Di Persio J. Therapy Options in Imatinib Failures. The Oncologist (2008). , 13, 424-434.

[77] Hare, O, Walters, T, & Stoffregen, D. E, et al. In vitro activity of Bcr-Abl inhibitors AMN107 and BMS 354825 against clinically relevant imatinib-resistant Abl kinase domain mutants. Cancer Research (2005). , 65, 4500-4505.

[78] Belloc, F. Moreau Gaudry F, Uhalde M, et al. Imatinib and nilotinib induce apoptosis of chronic myeloid leukemia cells through a Bim-dependeant pathway modulated by cytokines. Cancer Biology and Therapy (2007). , 6, 912-919.

[79] Golemovic, M, Verstovsek, S, Giles, F, et al. AMN 107, a novel aminopyrimidine inhibitor of Bcr-Abl, has in vitro activity against imatinib-resistan chronic myeloid leukemia. Clinical Cancer Research (2005). , 11, 4941-4947.

[80] Copland, M, Pellicano, F, Richmond, L, et al. BMS 214662 potently induces apoptosis of chronic myeloid leukemia stem and progenitor cells and synergizes with tyrosine kinase inhibithors. Blood (2008). , 111, 2843-2854.

[81] Woessner, D, Lim, C, & Deininger, M. Development of an effective therapy for CML. Cancer Journal (2011). , 17, 477-486.

[82] Jean Khoury HCortes J, Dantarjian H, et al. Bosutinib is active in chronic phase chronic myeloid leukemia after imatinib and dasatinib and/or nilotinib therapy failure Blood; (2012). , 119, 4303-4312.

[83] Hare, O, Shakespeare, T, & Zhu, W. X, et al. AP24534 a pan BCR-ABL inhibitor for chronic myeloid leukemia potently inhibits the T315I mutant and overcomes mutation based resistance. Cancer Cell (2009). , 16, 401-412.

[84] Balabanov, S, Gontarewicz, A, Keller, G, et al. Abcg2 overexpression represents a novel mechanism for acquired resistance to multi-kinase inhibithor danusertib in BCR-ABL positive cells in vitro. (2011). Plos One. 6: ee19164., 19146.

[85] Cortes, J, Jabbour, E, Kantarjian, H, et al. Dynamics of BCR-ABL kinase domain mutations in chronic myeloid leukemia after sequential treatment with multiple tyrosine kinase inhibitors. Blood (2007). , 110, 4005-4011.

[86] Kamitsuii, Y, Kuroda, J, Kimura, S, et al. The Bcr-Abl kinase inhibitor INNO 406 induces autophagy and different modes of cell death execution in Bcr-Abl positive leukemias. Cell Death and Differentiation (2008). , 11, 1712-1722.

[87] Jabbour, E, Hochhaus, A, Cortes, J, et al. Choosing the best treatment strategy for chronic myeloid leukemia patients resistant to imatinib: weighing the efficacy and safety of individual drugs with BCR ABL mutations and patient history. Leukemia (2010). , 24, 6-12.

[88] Baccarani, M, Cortes, J, Pane, F, et al. Chronic myeloid leukemia: an update of concepts and management recommendations of European Leukemia Net. Journal of Clinical Oncology (2009). , 27, 6041-6051.

[89] Ernst, T, & Hochhaus, A. Chronic Myeloid Leukemia: Clinical Impact of BCR-ABL1 Mutations and Other Lesions Associated With Disease Progression. Seminars in Oncology (2012). , 39, 58-66.

[90] Hochhaus, A. La Rosée P. Imatinib therapy in chronic myelogenous leukemia: strategies to avoid and overcome resistance. Leukemia (2004). , 18, 1321-31.

[91] De Lavallade, H, Apperley, J. F, Khorashad, J. S, et al. Imatinib for newly diagnosed patients with chronic myeloid leukemia: incidence of sustained responses in an intention-to-treat analysis. Journal of Clinical Oncology (2008). , 26, 3358-3363.

[92] Hughes, T, & Branford, S. Molecular monitoring of BCR-ABL as a guide to clinical management in chronic myeloid leukaemia. Blood Reviews (2006). , 20, 29-41.

[93] Ernst, T, Hoffmann, J, Erben, P, et al. ABL single nucleotide polymorphisms may masquerade as BCR-ABL mutations associated with resistance to tyrosine kinase inhibitors in patients with chronic myeloid leukemia. Haematologica (2008). , 93, 1389-1393.

[94] Gorre, M, Mohammed, M, Ellwood, K, et al. Clinical resistance to STI-571 cancer therapy caused by BCRABL gene mutation or amplification. Science (2001). , 293, 876-880.

[95] Hochhaus, A, Kreil, S, Corbin, A. S, et al. Molecular and chromosomal mechanisms of resistance to imatinib (STI571) therapy. Leukemia (2002). , 16, 2190-2196.

[96] Branford, S, Rudzki, Z, Walsh, S, et al. Detection of BCR-ABL mutations in patients with CML treated with imatinib is virtually always accompanied by clinical resistance, and mutations in the ATP phosphate- binding loop (P-loop) are associated with a poor prognosis. Blood (2003). , 102, 276-283.

[97] Chu, S, Xu, H, Shah, N. P, et al. Detection of BCR-ABL kinase mutations in CD34+ cells from chronic myelogenous leukemia patients in complete cytogenetic remission on imatinib mesylate treatment. Blood (2005). , 105, 2093-2098.

[98] Hare, O, Eide, T, Deininger, C, & Bcr-abl, M. kinase domain mutations, drug resistance, and the road to a cure for chronic myeloid leukemia. Blood (2007). , 110, 2242-2249.

[99] La Rosee PHochhaus A. Resistance to Imatinib in Chronic Myelogenous Leukemia: Mechanisms and Clinical Implications. Current Hematologic Malignant Reports (2008). , 3, 72-79.

[100] Cortes, J, Kim, D, & Pinilla-ibarz, J. Initial findings from the PACE trial: a pivotal phase 2 study of ponatinib in patients withCML and Ph + ALL resistant or intolerant to dasatinib or nilotinib, or with the T315I mutation. ASH Annual Meeting Abstract (2011).

[101] Diamond, J, & Melo, J. Mechanisms of resistance to BCR-ABL kinase inhibitors. Leukemia and Lymphoma (2011). S1):12-22.

[102] Weisberg, E, & Griffin, J. Mechanism of resistance to the ABL tyrosine kinase inhibitor STI571 in BCR/ABL transformed hematopoietic cell lines. Blood (2000). , 95, 3498-3505.

[103] Le Coutre PTassi E, Varella-Garcia M, et al. Induction of resistance to the Abelson inhibitor STI571 in human leukemic cells through gene amplification. Blood (2000). , 95, 1758-1766.

[104] Barnes, D, Palaiologou, D, Panousopoulou, E, et al. Bcr-Abl expression levels determine the rate of development of resistance to imatinib mesylate in chronic myeloid leukemia. Cancer Research (2005). , 65, 8912-8919.

[105] Uchida, N, Combs, J, Chen, S, et al. Primitive human hematopoietic cells displaying differential efflux of the rhodamine 123 dye have distinct biological activities. Blood (1996). , 88, 1297-1305.

[106] Chaudhary, P, & Roninson, I. Expression and activity of P glycoprotein, a multidrug efflux pump, in human hematopoietic stem cells. Cell (1991).

[107] Leslie, E, Deeley, R, & Cole, S. Multidrug resistance proteins: role of P glycoprotein, MRP1, MRP2 and BCRP(ABCG2) in tissue defense. Toxicology and Applied Pharmacology (2005). , 204, 216-37.

[108] Mahon, F. X, Deininger, M. W, Schultheis, B, et al. Selection and characterization of BCR-ABL positive cell lines with differential sensitivity to the tyrosine kinase inhibitor STI571: diverse mechanisms of resistance. Blood (2000). , 96, 1070-1079.

[109] Hamada, A, Miyano, H, Watanabe, H, et al. Interaction of imatinib mesilate with human P-glycoprotein. Journal of Pharmacology and Experimental Therapeutics (2003). , 307, 824-828.

[110] Illmer, T, Schaich, M, Platzbecker, U, et al. P-Glycoprotein-mediated drug efflux is a resistance mechanism of chronic myelogenous leukemia cells to treatment with imatinib mesylate. Leukemia (2004). , 18, 401-408.

[111] Burger, H, Van Tol, H, & Brok, M. Chronic imatinib mesylate exposure leads to reduced intracellular drug accumulation by induction of the ABCG2 (BCRP) and ABCB1 (MDR1) drug transport pumps. Cancer Biology and Therapy (2005). , 4, 747-752.

[112] Jordanides, N. E, Jorgensen, H. G, Holyoake, T. L, et al. Functional ABCG2 is over expressed on primary CML CD34+ cells and is inhibited by imatinib mesylate. Blood (2006). , 108, 1370-1373.

[113] Burger, H, & Nooter, K. Pharmacokinetic resistance to imatinib mesylate: role of the ABC drug pumps ABCG2 (BCRP) and ABCB1 (MDR1) in the oral bioavailability of imatinib. Cell Cycle (2004). , 3, 1502-1505.

[114] Brozik, A, Hegedus, C, Erdei, Z, Hegedus, T, Ozvegy-laczka, C, & Szakacs, G. Tyrosine kinase inhibitors as modulators of ATP binding cassette multidrug transporters: substrates, chemo-sensitizers or inducers of acquired multidrug resistance?. Expert Opinion on Drug Metabolism and Toxicology (2011). , 7, 623-42.

[115] Thomas, J, Wang, L, Clark, R. E, et al. Active transport of imatinib into and out of cells: implications for drug resistance. Blood (2004). , 104, 3739-45.

[116] Crossman, L, Druker, B, & Deininger, M. hOCT 1 and resistance to imatinib. Blood (2005). , 106, 1133-1134.

[117] White, D, Saunders, V, & Dang, P. CML patients with low OCT-1 activity achieve better molecular responses on high dose imatinib than on standard dose. those with high OCT-1 activity have excellent responses on either dose: a TOPS correlative study. Blood (2008).

[118] Engler, J, Frede, A, Saunders, V, et al. Chronic Myeloid Leukemia CD34+ cells have reduced uptake of imatinib due to low OCT-1 activity. Leukemia (2010). , 24, 765-770.

[119] White, D, Saunders, V, Dang, P, et al. OCT-1-mediated influx is a key determinant of the intracellular uptake of imatinib but not nilotinib (AMN107): reduced OCT-1 activity is the cause of low in vitro sensitivity to imatinib. Blood (2006). , 108, 697-704.

[120] Hiwase, D, Saunders, V, Hewett, D, et al. Dasatinib cellular uptake and efflux in chronic myeloid leukemia cells: therapeutic implications. Clinical Cancer Research (2008). , 14, 3881-3888.

[121] Wang, J, Lapidot, T, Cashman, J, et al. High level engraftment of NOD/SCID mice by primitive normal and leukemic hematopoietic cells from patients with chronic myeloid leukemia in chronic phase. Blood (1998). , 91, 2406-2414.

[122] Holyoake, T. L, Jiang, X, Jorgensen, H. G, et al. Primitive quiescent leukemic cells from patients with chronic myeloid leukemia spontaneously initiate factor-independent growth in vitro in association with up-regulation of expression of interleukin-3. Blood (2001). , 97, 720-728.

[123] Copland, M, Hamilton, A, Elrick, L. J, et al. Dasatinib (BMS-354825) targets an earlier progenitor population than imatinib in primary CML but does not eliminate the quiescent fraction. Blood (2006). , 107, 4532-4539.

[124] Jiang, X, Zhao, Y, Chan, W. Y, et al. Leukemic stem cells of chronic phase CML patients consistently display very high BCR-ABL transcript levels and reduced responsiveness to imatinib mesylate in addition to generating a rare subset that produce imatinib mesylate resistant differentiated progeny. Blood (2004). a.

[125] Keeshan, K, Mills, K, Cotter, T, et al. Elevated Bcr-Abl expression levels are sufficient for a haematopoietic cell line to acquire a drug-resistant phenotype. Leukemia (2001).

[126] Sawyers, C. L. Signal transduction pathways involved in BCR-ABL transformation. Baillieres Clinical Haematology (1997). , 10, 223-231.

[127] Wu, J, Meng, F, Kong, L, et al. Association between imatinib-resistant BCR-ABL mutation negative leukemia and persistent activation of LYN kinase. Journal of the National Cancer Institute (2008). , 100, 926-939.

[128] Mein, M. A. rd, Wilson M, Abdi F, et al. Src family kinases phosphorylate the Bcr-Abl SH3-SH2 region and modulate Bcr-Abl transforming activity. Journal of Biological Chemistry (2006). , 281, 30907-30916.

[129] De Groot, R, Raaijmakers, J, Lammers, J, et al. STAT5 activation by BCR-Abl contributes to transformation of K562 leukemia cells. Blood (1999). , 94, 1108-1112.

[130] Quintas-cardama, A, & Cortes, J. Molecular biology of bcr-abl1-positive chronic myeloid leukemia. Blood (2009). , 113, 1619-30.

[131] Pendergast, A, Quilliam, L, Cripe, L, et al. BCR/ABL- induced oncogenesis is mediated by direct interaction with the SH2 domain of the GRB-2 adaptor protein. Cell (1993). , 75, 175-185.

[132] Sattler, M, Mohi, M, Pride, Y, et al. Critical role for Gab2 in transformation by BCR/ABL. Cancer Cell (2002). , 1, 479-492.

Chapter 9

Role of Bone Marrow Derived Mesenchymal Stem Cells in Management of Graft Versus Host Disease

Aisha Nasef

Additional information is available at the end of the chapter

http://dx.doi.org/10.5772/55013

1. Introduction

Human bone marrow stromal cells, referred as mesenchymal stem cells (MSC) are multipotent unspecialized cells localized in the medullary stroma [1,2]. They have a capacity for self-renewal and differentiation into multiple cell lineages [3-6].

Mesenchymal and tissue stem cell committee of the International society for cellular therapy proposes minimal criteria to define human MSCs. First, MSCs must be plastic-adherent when maintained in standard culture conditions. Second, MSCs must express CD105, CD73 and CD90, and lack the expression of CD45, CD34, CD14 or CD11b, CD79alpha or CD19 and HLA-DR surface molecules. Third, MSCs must be capable to differentiate into osteoblasts, adipocytes and chondroblasts in vitro [7,8]. In the hand of Delorme B and Charbord P, MSCs can be defined phenotypically with a minimal set of markers as CD31-, CD34-, and CD45-negative cells and CD13, CD29, CD73, CD90, CD105, and CD166 positive cells [9]. MSCs have been used in cell-based therapy in various disease conditions. Experimental evidence and preliminary clinical studies have demonstrated that MSCs, have an important immunomodulatory function in the management of allogeneic hematopoietic stem cell (HSC) transplantation [10,11]. These immunomodulatory effects have been demonstrated in various species, including humans, rodents, and primates and show clinical promise for the treatment of graft versus host disease (GVHD) and graft failure management. In this chapter we will discuss current research finding.

2. General background

2.1. MSC administration

2.1.1. Safety and therapeutic methods of MSC administration

The extensive capacity for expansion in vitro at clinical scale has recently facilitated the development of clinical trials designed to assess safety, feasibility, and efficacy of transplanting MSCs for a variety of diseases [11]. There was no toxicities related to the infusion of expanded autologus MSCs into patients with advanced breast cancer or with haematological malignancy, into hurler syndrome patients and into patients of metachromatic leukodystrophy [12,13].

MSC could be delivered systematically and locally. Systemic delivery circumvent the problems such as tissue damage and unsuitability of delivering multiple doses. Site specific delivery has the advantage of delivering large numbers of cell directly to the required sites. Majority of studies showed that MSC infused systematically homed to injured tissues [14,15].

Successful systemic delivery of MSC is dependent upon efficient homing of the cells to the required sites. In this respect, migration of MSC from the circulation into damaged tissues leading to therapeutic effect s has been documented [16]. Ability of MSC to home to bone marrow (BM) can be affected by in vitro culture, which could be due to decrease of adhesion molecules and Chemokine receptors [17].

2.2. Mechanism of homing of MSC

Homing of leukocytes to inflammatory sites involve selectin, chemokines, integrins and other adhesion molecules [18]. Hemopoietic stem cells (HSC) recruited from blood vessels on similar process to that of leucocytes [19]. MSC recruited to damaged tissues. It is a fair assumption to suppose that they utilize comparable mechanisms of recruitment of MSC. P-selectin play an important role in the trafficking of MSC [20]. MSC roll upon endothelial cells as the first stage in their recruitment. Chemokine receptors are expressed on the cell surface of MSC [21] and their stimulation has been shown to induce cell migration and directing MSC.

2.3. Transplant ability, engraftment and tracking of MSC

Numerous studies have demonstrated migration and multiorgan engraftment of MSCs both in animal models and in human clinical trials. [22-24].

Direct injection of human marrow stromal cells into the corpus striatum of rat brain showed engraftment of 20% of the infused cells [25]. Rat bone marrow stromal cells infused into briefly distally occluded ascending aorta migrated after 8 weeks in the scar and periscar tissue [26]. MSCs injected intravenously into irradiated primate had the ability of engraftment in different injured tissues as the bone marrow, skin, digestive tract and muscle [9,11]. In rat model, MSC engrafted in multiple organs such as lung, liver, kidneys and spleen.

Human MSCs engrafted and demonstrated site-specific differentiation in sheep [27,28] and in murine models [29-31]. The capacity of engrafment of MSCs was not influenced neither by the route of administration nor by the difference in conditioning protocols [32]. Clinically, both autologous or allogeneic MSCs were given to patients [33, 34]. Allogeneic HLA-mismatched male fetal MSCs injected into HLA-incompatible female fetus with osteogenesis imperfecta (OI) engrafted and differentiated into bone [35]. Haploidentical MSCs had a low level of engrafment in a patient with aplastic anemia, however there was a partial restoration of bone marrow microenvironement [36]. In contrast, Infused allogeneic MSCs did not expand significantly in patients and they originate from the host [34, 37]. It is of interest to know into which organs MSCs home. Allogeneic and autologus MSCs distributed to a wide range of tissues in baboons [22].

2.4. Transformation of MSCs

MSCs transformed in vitro. The transformation of MSCs is associated with chromosomal abnormalities, increased levels of telomerase activity and c-myc expression [38]. Human MSCs isolated from adipose tissue undergo spontaneous transformation after long-term culture (4-5months) [39]. Others found that short-term culture was sufficient for the transformation of murine MSCs into a cell population with autonomous growth and biological characteristics of osteosarcoma [40]. In contrast, even long term cultures could not induce MSC transformation [41]. Previous study reported that gastric cancer could originate from BM-derived cells, presumably MSCs [42]. Human BM-MSCs cultured extensively, with a high Telomerase activity is capable of forming solid tumors in multiple organs in mice [43].

MSCs could migrate towards primary tumors and metastatic sites. Chemokines secreted by MSCs have been shown to enhance the emergence of pulmonary metastases and such secretion has a strong interaction between breast cancer cells and MSCs. In addition, MSCs have also been found to play a role in drug resistance in various cancer cells. MSCs protect chronic lymphocytic leukemic cells from fludarabine-, dexamethosone-, and cyclophosphamide-induced apoptosis. MSC non-selectively protected chronic myeloid leukemia cells from imatinib-induced apoptosis. [44].

Indeed, the transformation potential in cultured MSCs must be documented before considering infusion of these cells into patients. However, this issue remains controversial, as other studies did not observe transformation of human BM-MSCs.

In conclusion, it is possible that the way MSCs are expanded and long term culture lead to transformation. The safety of using MSCs in humans remains open. The use of MSCs in patients should definitely require precise and limited procedures of expansion to avoid the risk of injecting transformed cells. These finding emphasize the need for accurate studies aimed at investigating the bio safety of these cells

2.5. MSCs and metastasis

So far, very few studies have addressed the question of the effects of MSCs on metastasis. A few studies have supported that MSCs may suppress tumor growth while others believe that MSCs may contribute to tumor protection via antiapoptotic effect, tumor progression, metastasis, and drug-resistance of cancer cells.

Human MSCs played a dual role in tumor cell growth *in vitro* and *in vivo*. It was found that human MSCs inhibited the proliferation of cancer cell lines and caused G1 phase cell cycle arrest and apoptosis *in vitro*. However they enhance tumor formation and growth *in vivo*. MSCs have also been found to prevent apoptosis of acute myeloid leukemia cells by up-regulation of antiapoptotic proteins [44].

The main adverse role of MSCs was its pro-invasive potential [45]. It is likely that other molecules participate to the enhancement of metastasis by MSCs and this will be the challenge of future studies.

To use MSCs in anti-cancer therapies, it will be essential to identify the factors produced by MSCs cells responsible for the inhibition or the enhancement of tumor growth and those governing the response of tumor cells.

Evaluation of the potential use of MSCs in cell-based anti-cancer therapies is just starting. These cells have shown some promise as several studies have reported that a portion (which remains to be defined) of MSCs is able to migrate to the tumor site. However, this homing of MSCs is not selective and it will be important to evaluate possible side effects in organs, which are not affected by the disease. Overall, MSCs represent great hope for cancer therapies, but a thorough evaluation of their potential risk will be pre-required step [46].

3. Immunological characteristics of MSCs

Studies in animal models and in humans demonstrated that co-transplantation of HSCs along with MSCs obtained by ex-vivo expansion enhance hematopoietic reconstitution [47,48]. Experimental and clinical studies demonstrated the safety and immunosuppressive role of MSCs infusion [49, 50].

3.1. Immunological characteristics of MSCs *in vivo*

It has been demonstrated since 1984 that reconstitution of irradiated host with T-cell depleted bone marrow containing both host (syngeneic) and donor (allogeneic or Xenogeneic) components leads to long-term survival of the reconstituted animals and specific prolongation of subsequent skin grafts of donor type. However these animals are fully reactive to third-party allograft and do not appear to manifest signs of graft-versus-host disease (GVHD) [51]. The possibility of the presence of immunomodulatory cells in bone has been noticed after donor specific long-term hypo responsive status obtained by transplantation of bone and HSCs in murine models [52, 53]. MSCs constitute the stromal scaffold which is close to the endostenum and interact tightly with HSCs [54]. They contribute to the formation of HSCs niche, support HSCs engraftment and survival of blood cells in vivo [11]. This indicate the presence of underlying immunomodulatory effect of MSCs.

3.1.1. *MSCs immunosuppressive effect in experimental animal models*

Various animal models have been used to study the immunomodulatory effects of MSCs in treatment of GVHD, autoimmune disorders and tumor immunity. It seems to be that early and repeated injection of MSCs following HSCs transplantation is primordial to control GVHD [55, 56]. In autoimmunity model only early and re-injection of MSCs at the peak of the disease were effective as compared to injection after disease stabilization [57]. MSCs do not elicit immunological reaction in recipients [49, 56]. MSCs play tolerogeneic effect in recipients [58-60]. However, MSCs had been rejected by mismatched recipients [61] and failed to prevent GVHD [56].

3.1.2. *MSCs immunosuppressive effect in humans*

In the field of HSC transplantation, there are two applications of MSCs:

- Improvement of stem cell engrafting and the acceleration of hematopoietic reconstitution based on the hematopoesis-supporting ability.
- Treatment of severe GVHD based on the immunomodulatory ability.

3.1.2.1. *Role of MSCs in support of hematopoiesis*

MSCs constitute the functional and structural support of medullary hematopoiesis by providing growth factors and extracellular matrix [62, 63]. MSCs have a positive impact on hematopoesis and results in rapid hematopoietic recovery [64].

Co transplantation of HSCs and haplotype-mismatched MSCs to patients with high-risk acute myelogenous leukemia result in rapid engraftment [65]. Co transplantation of allogeneic MSCs and multidonor umbilical cord blood (UCB) correlated with a higher overall level of engraftment into NOD/SCID mice [66]. In European phase I-II study co-infusion of haploidentical HSCs and MSCs accelerated leukocyte recovery as compared to control group [67]. In multicentre clinical trial, co transplantation of MSCs with an HLA-identical sibling HSCs after a myeloablative conditioning regimen induced hematopoietic recovery of peripheral mono nuclear cells (MNC) and platelets [68, 69] and resulted in fast engraftment of absolute neutrophils count and 100% donor chimerism [69]. Transplantation of MSCs into immunosupressed patients generated neither alloantibody against MSC nor against fetal calf serum (FCS) [70].

3.1.2.2. *Role of MSCs in management of graft versus host disease (GVHD)*

Injection of MSCs could cure severe graft versus host disease (GVHD) and promote hematopoietic recovery.

MSC-mediated inhibition of immune response is a complex mechanism involving changes in the maturation of antigen-presenting cells and in cytokine secretion profiles as well as the suppression of monocyte differentiation into dendritic cells [71]. They exert profound immunosuppression by inhibiting T-cell proliferation in response to various stimuli in vitro. They induce regulatory immunosuppressive lymphocytes and CD8 apoptosis. MSCs inhibit cell cycle progression and CD4 allo-proliferation. This immunosuppressive effect of MSCs is mediated through several inducible soluble factors [71].

3.1.2.3 Clinical trials of GVHD management by MSCs

Patients treated with MSCs had less GVHD and less toxicity in a retrospective study comparing allograft of geno-identical HSCs with and without MSCs [72]. No acute side effects occurred after the infusion of haploidentical and mismatched MSCs from unrelated donors into patients suffering from GVHD [50, 73]. Haploidentical MSCs was used to treat grade IV GVHD of the gut and liver that was resistant to conventional treatment. The aim was to use the tissue repair effect shown in vivo in animal models, and the immunomodulatory effects seen in vitro on human lymphocytes. The clinical response was striking with normalization of stool and bilirubin on two occasions. They leads to healing of damaged bowel epithelia [50]. The patient was highly immunosupressed. However, his lymphocytes continued to proliferate in response to lymphocytes of MSCs donor in vitro in several occasions. This could indicate an immunosuppressive effect of MSCs rather than tolerance induction. We would be able to prevent GVHD after MSCs infusion and maintaining in the same time the response of host lymphocytes against alloantigens. Infusion of MSCs into 8 patients with steroid refractory GVHD cause dramatical disappearance of all symptoms and repaired gut in six patients and liver in one. The survival curve was better than that of 16 patients with steroid-resistant biopsy-proven gastrointestinal GVHD, not treated with MSCs [245]. MSCs were used to treat 40 patients with acute and chronic GVHD. More than forty-seven (47.5 %) showed complete response, 22.5 % showed improvement, 10 % had stable disease and 17.5 % had no response. Between 6 weeks up to 3.5 years after transplantation more than half of patients are alive [73].

At the same time of hematopoietic stem cell transplantation (HSCT), 46 patients received culture-expanded MSCs from their HLA-identical sibling donors [74]. Moderate to severe acute GVHD was observed in 13 (28%) of 46 patients. Chronic GVHD was observed in 22 (61%) of 36 patients and 2-year survival was 53%. No MSC-associated toxicities were seen. Stromal cell chimerism was demonstrated in 2 of 19 examined patients at 6 and 18 months after transplantation. MSCs are safe to give, but are difficult to detect after infusion, even in immunocompromised patients who have undergone HSCT [75].

Ten patients undergoing HSCT were treated with MSCs due to tissue toxicity. In five patients, severe hemorrhagic cystitis cleared after MSC infusion. Gross hematuria disappeared after a median of 3 days. Two patients with grade five hemorrhagic cystitis had reduced transfusion requirements after MSC infusions, but both died of multi-organ failure. MSC donor DNA was demonstrated in the urinary bladder in one of them. Two patients were treated for pneumomediastinum, which disappeared after MSC infusion. A patient with steroid-resistant GVHD of the gut experienced perforated diverticulitis and peritonitis that was dramatically reversed twice after infusion of MSCs[76].

In Europe, 55 patients have been treated for steroid-resistant acute GVHD with an overall response rate of 69%. Non responders have died of progressive GVHD and several responders have died from infections with an overall survival of 23 of 55 (42%) from 2 months to 5 years. Although the experience is limited, MSCs seems a promising treatment for severe steroid-resistant acute GVHD [73].

In a phase I/II clinical trial, ten percent of patients treated for acute refractory GVHD obtained a complete response, 60 % had a partial response and 30 % did not respond. They found that 50% of patients with chronic refractory GVHD did not respond, 12.5 % had complete remission and 37.5% had partial response [77]. Transplantation of MSCs into 15 patients with haematological malignancies was safe and induced complete remission [78].

These preliminary data suggest that MSCs may also play a role in repairing severe tissue toxicity. It seems to be that there is beneficial effect of MSCs infusion in humans. However, The optimal MSCs dose and frequency of administration needed to be evaluated to control GVHD.

The underlying molecular mechanisms of immunosuppression in vivo are unknown. It has been demonstrated that MSCs engrafted to injured tissues rich on inflammatory cytokines [107]. MSCs inhibitory effect is inducible [79] and presence of MSCs in such media could explain partially the beneficial effect of MSCs infusion in the recipients, despite non detectable MSCs in recipients BM.

3.2. Suggested mechanisms for MSC immunomodulatory effect

Co-culture of MSCs with allogeneic lymphocytes failed to stimulate their proliferation, indicating that these cells are innately not immunogeneic. Recent reports suggest that MSCs have immunomodulatory properties and can inhibit lymphocyte antigen presenting cells, natural killer cells, and cytotoxic lymphocyte proliferation in mixed-lymphocyte reactions (MLR). MSCs inhibit CD2, CD4 and CD8 subsets of T lymphocytes. The immunosuppressive effect of human MSCs was higher in cultures with cell contact than in cultures without contact (transwell), and the difference was statistically significant [80, 81]. MSCs produce a variety of growth factors that are likely to play a role in immunomodulation. Human and murine MSCs do inhibit the proliferation of lymphocytes in MLR by soluble factors [80-83]. Indoleamine 2,3 dioxygenase (IDO) inhibit the alloreactivity induced by antigen presenting cells (APC) [84]. The inhibitory effect of MSCs on alloreactivity seems to be due to other mechanisms rather than apoptosis of lymphocytes. Several studies demonstrated reversibility of MSCs inhibitory effect. [81, 85, 86].

Cell anergy is a state of immune unresponsiveness, defines the inability of an immune cell to mount a complete response against its target. MSCs induce anergy due to divisional arrest of T cells. IFN-γ production but not proliferation of murine cells was restored [87]. Restoration of CD4+ cell proliferation but not CD8+ cells after the removal of human MSCs was observed. For the time being we could assume that MSCs do induce anergy [88]. MSCs induced regulatory cells [88]. Regulatory cells have been involved in the regulation of immune response.

Despite the expression of human leukocyte antigen (HLA) by MSCs, they were well tolerated without side effects in allogeneic hosts. Major Histocomaptability Complex (MHC) had no role in MSCs inhibitory effect as autologus and allogenic human [81], murine [89] and baboon [49]. MSCs were capable of inhibiting the proliferation of lymphocytes. In addition, Xenogeneic murine MSCs inhibited the proliferation of human cells [79]. However, one study showed that transplantation of allogeneic MSCs resulted in rejection by class I and class II mismatched recipients in murine model [61].

The major effectors of the innate immunity are natural killer (NK) cells. They eliminate malignant and virus infected cells. MSCs alter the phenotype of NK cells and suppress proliferation and cytotoxicity against HLA-class I-expressing targets in time and dose dependent manner in cell contact or transwell cultures [90, 91].

Veto activity was defined by Miller [92] as the ability to induce specific suppression of cytotoxic T lymphocyte precursors (CTL-Ps) against antigens present on veto cell surface, but not against those on third-party allogeneic cells. MSCs had veto activity [93]. This is contradictory to all the present data where MSCs were able to inhibit the allo-response to allogeneic lymphocytes in MLR in different species [79, 81]. Even more, veto property contrasts with the proliferation of allogeneic lymphocytes observed in the presence of MSCs by the same team [93].

There was drastic reduction of the recipient's CTL response against injected class I antigens due to veto phenomenon [94]. These findings still keep on with Miller's definition [92]. Veto mechanism could play a role in low immunogenicity of MSCs, but certainly doesn't explain the suppression of allogenic lymphocytes by MSCs [71]. MSCs have no effect on the lymphocyte response to recall antigen [93].

Human MSC inhibited the pro--inflammatory Th1 cytokines (IFN-γ, TNF- α, IL-1α and IL-β) whereas the anti-inflammatory Th2 cytokines IL-3, IL-5, IL-10, IL-4 and IL-13 and the Th2 Chemokine I-309 (a chemo attractant for regulatory T cells) were increased [71, 95]. This could indicate a shift from the prominence of pro-inflammatory Th1 cells towards an increase in anti-inflammatory TH2 cells and support MSC inhibitory properties. MSCs can skew the dendritic cells (DCs) function, thus biasing the immune system toward Th2 and away from Th1 responses [96]. MSC had no effect on neutrophils phagocytosis, expression of adhesion molecules and chemotaxis [97]. This is an interesting finding, as MSCs could not interfere with neutrophils function. Summary of MSCs inhibitory effect is illustrated in figure-1. For further reading, about detailed description of possible MSCs inhibitory mechanisms, refer to this article [10].

3.2.1. Inhibitory effect of autogenic and allogenic MSCs

Allogeneic MSCs have stronger immunosuppressive effects than autologus MSCs [88]. Others found comparable and significant inhibition was elicited by autologus or allogeneic MSCs [81,85]. This could help in management of GVHD in cases where non-matched HSCs are used

4. What starts the first step in induction of inhibitory effect in MSC/Mixed lymphocyte reaction MLR?

It is not known for the time being whether MSCs or lymphocytes do start the first signals to activate the other one to induce the immunosuppressive effect. MSC supernatant failed to

http://dx.doi.org/10.5772/55013

inhibit T-cell activation and, on the contrary, MSC supernatant had a stimulatory effect on MLR. Surprisingly, cell-free supernatant obtained from MLR had an inhibitory effect similar to that seen with the addition of MSCs [58]. In agreement Djouad et al. found that only MSC conditioned supernatant by lymphocytes was capable to inhibit secondary MLR but less efficiently than the co-culture of cell partners, in indicating that the immunosuppressive properties of MSCs are mediated by inducible soluble factors and the interaction between the MSCs and lymphocytes is a pre-requirement for MSC mediated inhibitory effect [79].

In addition, an enhancement of the MSC immunosuppressive effect was observed after addition of irradiated third party MNC to MSC culture, indicating that the physical interaction between MSC and PBMC increase the suppressive activity [98]. Human MSCs do not constitutively express IDO. But activation by IFN-γ secreted by allogeneic lymphocytes in a dose dependent manner is required for induction of functional IDO activity [99].

It has been demonstrated that auto reactive cells may induce the transdifferentiation of MSCs to neural stem cells. This phenomenon could be due to stimulation of cytokine production and generation of suitable environment that results in differentiation into neural stem cells [100]. All these data could confirm the interaction between MSCs and lymphocytes. Physical interaction between MSCs and T-cells increases the suppressive activity as there was an increased expression of IL-10 and TGF-β genes [80] as compared to non contact cultures. All these data could complicate the issue. Is MSC mediated inhibitory effect is a consequence of MSC activation by IFN-γ secreted by two allogeneic populations of lymphocytes in MLR?. Why there is no stimulation of single allogeneic PBL by allogeneic MSCs?

Are there two separate effector mechanisms, one to escape the immunological recognition and the other to inhibit the alloreactivity? Is the constitutive expression of TGF-β and HLA-G by the MSCs play a role in immunological escape?

Maccario et al. had demonstrated the absence of regulatory cell in co-culture of MSCs with single allogeneic PBMCs as compared to the presence of MSCs in MLR [88]. This could indicate that allogeneic lymphocytes initiate the first signals to stimulate the production of molecules that induce the production of regulatory cells.

It has been found that CD14+ monocyte are the PBMC subpopulation, being responsible for MSC activation through IL-B secretion [101]. MSC inhibitory effect was mediated through CD8+ in human and murine model [97, 88]. CD8+ cells are the executive cells for MSC inhibitory effect rather than being the inducers of MSC inhibitory effect. It has been demonstrated that MSC inhibitory mechanisms differ depending on nature of stimulus [102]. This could indicate that alloantigens or mitogen are responsible for the first stimulating signals in MLR, or at least they are responsible for later on modification of these signals. It has been found that lymphocytes and MSCs are mutually inhibitory on their respective proliferation and indicate the bi-directional interaction and cross talk between lymphocytes and MSCs [103, 104]. More recently, it has been found that IFNγ and concomitant presence of TNFα or IL-1α or IL-1β induce the expression of several chemokine and inducible Nitric oxide synthase by MSCs [105].

This could indicate that lymphocytes start the first signals for MSCs activation, however cross talk between both of them are essential to have the full immunological effects.

In conclusion, MSC must be handled with extreme caution before a large scale clinical trial is performed. it has been found in a pilot clinical study, that co transplantation of MSC and HSC prevent GVHD, but caused a higher relapse rate in hematologic malignancy patients as compared to control [106].

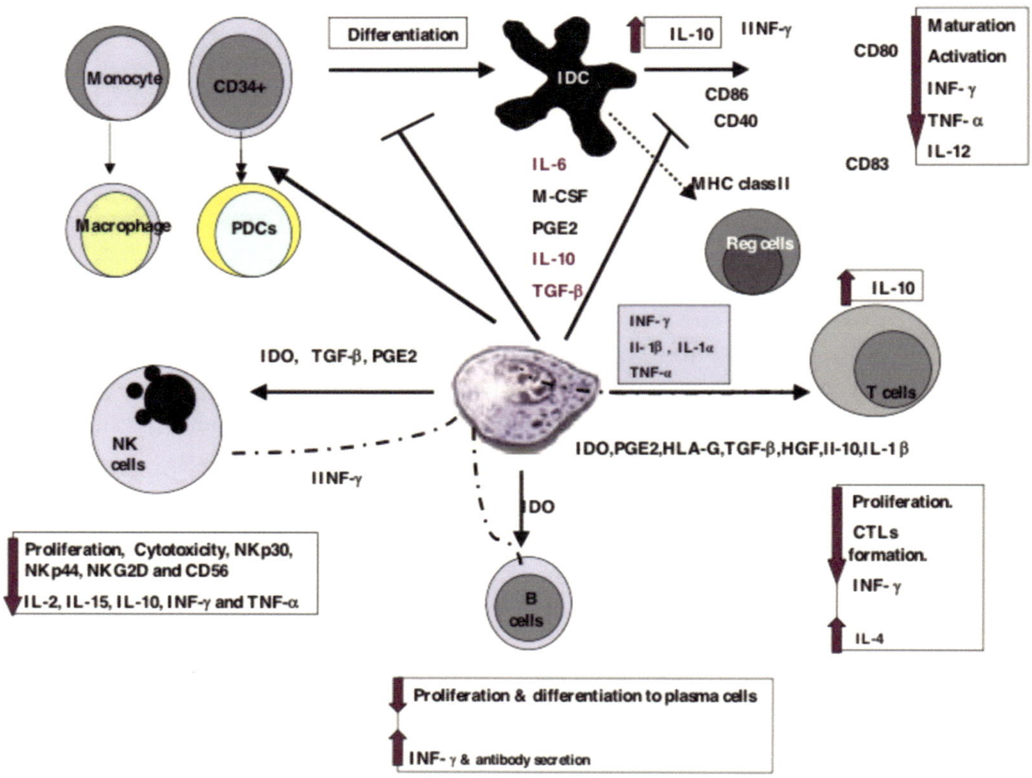

Figure 1. Mechanisms of action of MSCs upon dendrietic, lymphocytes T, lymphocytes B and natural killer cells.

MSCs inhibit monocytes (MO) differentiation into immature DCs (IDC) and skewed them toward macrophage. MSCs reduced percentage of CD34(+) derived IDC and increased plasmacytoid DC (pDC). They inhibit maturation IDC into mature DCs. They inhibit the regulation of HLA-DR, CD80, CD86, CD1a, CD40, CD14 and CD83 antigen on DCs surface through IL-6, M-CSF, PGE2, IL-10 and TGF- b secretion. MSCs increased production of IL-10 and decreased production of Il-12, TNF-a and INF-g by DCs. MSCs induced DC that exhibit a suppressor phenotype (supp APC) and generated alloantigen-specific regulatory cells (Reg cells). MSCs inhibit NK proliferation and change cytokines secretion pattern through IDO, TGF-b, PGE2. MSCs inhibit lymphocytes B Proliferation & differentiation to plasma cells

http://dx.doi.org/10.5772/55013

through IDO. MSCs inhibit lymphocytes T proliferation and cytotoxicity through IDO, PGE2, HLA-G, TGF-b, HGF, Il-10 and IL-1 b.

Author details

Aisha Nasef*

Address all correspondence to: Nasef@doctor.com

National Medical Research Centre NMRC, Libya

References

[1] Friedenstein, A. J, Chailakhyan, R. K, Latsinik, N. V, Panasyuk, A. F, & Keiliss-borok, I. V. Stromal cells responsible for transferring the microenvironment of the hemopoietic tissues. Cloning in vitro and retransplantation in vivo. Transplantation (1974)., 17(1974), 331-40.

[2] Friedenstein, A. J. Precursor cells of mechanocytes. Int Rev Cytol (1976)., 47(1976), 327-59.

[3] Friedenstein, A. J, Petrakova, K. V, Kurolesova, A. I, & Frolova, G. P. Heterotopic of bone marrow.Analysis of precursor cells for osteogenic and hematopoietic tissues. Transplantation (1968)., 6(1968), 230-47.

[4] Friedenstein, A. J, Chailakhyan, R. K, & Gerasimov, U. V. Bone marrow osteogenic stem cells: in vitro cultivation and transplantation in diffusion chambers. Cell Tissue Kinet (1987)., 20(1987), 263-72.

[5] Owen, M, & Friedenstein, A. J. Stromal stem cells: marrow-derived osteogenic precursors. Ciba Found Symp (1988)., 136(1988), 42-60.

[6] Prockop, D. J. Marrow stromal cells as stem cells for nonhematopoietic tissues. Science (1997)., 276(1997), 71-4.

[7] Horwitz, E. M, Le, K, Blanc, M, Dominici, I, Mueller, I, Slaper-cortenbach, F. C, Marini, R. J, & Deans, D. S. Krause, and A. Keating, Clarification of the nomenclature for MSC: The International Society for Cellular Therapy position statement. Cytotherapy (2005)., 7(2005), 393-5.

[8] Dominici, M, Le, K, Blanc, I, Mueller, I, Slaper-cortenbach, F, Marini, D, Krause, R, Deans, A, & Keating, D. Prockop, and E. Horwitz, Minimal criteria for defining multipotent mesenchymal stromal cells. The International Society for Cellular Therapy position statement. Cytotherapy (2006)., 8(2006), 315-7.

[9] Delorme, B, & Charbord, P. Culture and characterization of human bone marrow mesenchymal stem cells. Methods Mol Med (2007). , 140(2007), 67-81.

[10] Lazarus, H. M, Koc, O. N, Devine, S. M, Curtin, P, Maziarz, R. T, Holland, H. K, Shpall, E. J, Mccarthy, P, Atkinson, K, Cooper, B. W, Gerson, S. L, Laughlin, M. J, Loberiza, F. R, Jr, A. B, & Moseley, A. Bacigalupo, Cotransplantation of HLA-identical sibling culture-expanded mesenchymal stem cells and hematopoietic stem cells in hematologic malignancy patients. Biol Blood Marrow Transplant (2005). , 11(2005), 389-98.

[11] Nasef, A, Ashammakhi, N, & Fouillard, L. Immunomodulatory Effect of Mesenchymal Stem Cells: Possible mechanisms. Regenerative medicine ((2008). Review article..

[12] Koc, O. N, Gerson, S. L, Cooper, B. W, Dyhouse, S. M, Haynesworth, S. E, Caplan, A. I, & Lazarus, H. M. Rapid hematopoietic recovery after coinfusion of autologous-blood stem cells and culture-expanded marrow mesenchymal stem cells in advanced breast cancer patients receiving high-dose chemotherapy. J Clin Oncol (2000). , 18(2000), 307-16.

[13] Koc, O. N, Day, J, Nieder, M, Gerson, S. L, Lazarus, H. M, & Krivit, W. Allogeneic mesenchymal stem cell infusion for treatment of metachromatic leukodystrophy (MLD) and Hurler syndrome (MPS-IH). Bone Marrow Transplant (2002). , 30(2002), 215-22.

[14] Ramasamy, R, Lam, E. W, Soeiro, I, Tisato, V, Bonnet, D, & Dazzi, F. Mesenchymal stem cells inhibit proliferation and apoptosis of tumor cells: impact on in vivo tumor growth. Leukemia (2007). , 21(2007), 304-10.

[15] Francois, S, Bensidhoum, M, Mouiseddine, M, Mazurier, C, Allenet, B, Semont, A, Frick, J, Sache, A, Bouchet, S, Thierry, D, Gourmelon, P, Gorin, N. C, & Chapel, A. Local irradiation not only induces homing of human mesenchymal stem cells at exposed sites but promotes their widespread engraftment to multiple organs: a study of their quantitative distribution after irradiation damage. Stem Cells (2006). , 24(2006), 1020-9.

[16] Chen, J, Li, Y, Wang, L, Lu, M, Zhang, X, & Chopp, M. Therapeutic benefit of intracerebral transplantation of bone marrow stromal cells after cerebral ischemia in rats. J Neurol Sci (2001). , 189(2001), 49-57.

[17] Rombouts, W. J, & Ploemacher, R. E. Primary murine MSC show highly efficient homing to the bone marrow but lose homing ability following culture. Leukemia (2003). , 17(2003), 160-70.

[18] Springer, T. A. Traffic signals for lymphocyte recirculation and leukocyte emigration: the multistep paradigm. Cell (1994). , 76(1994), 301-14.

[19] Chute, J. P. Stem cell homing. Curr Opin Hematol (2006). , 13(2006), 399-406.

[20] Ruster, B, Gottig, S, Ludwig, R. J, Bistrian, R, Muller, S, Seifried, E, Gille, J, & Henschler, R. Mesenchymal stem cells display coordinated rolling and adhesion behavior on endothelial cells. Blood (2006). , 108(2006), 3938-44.

[21] Fox, J. M, Chamberlain, G, Ashton, B. A, & Middleton, J. Recent advances into the understanding of mesenchymal stem cell trafficking. Br J Haematol (2007). , 137(2007), 491-502.

[22] Devine, S. M, Cobbs, C, Jennings, M, Bartholomew, A, & Hoffman, R. Mesenchymal stem cells distribute to a wide range of tissues following systemic infusion into nonhuman primates. Blood (2003). , 101(2003), 2999-3001.

[23] Li, Y, Hisha, H, Inaba, M, Lian, Z, Yu, C, Kawamura, M, Yamamoto, Y, Nishio, N, Toki, J, Fan, H, & Ikehara, S. Evidence for migration of donor bone marrow stromal cells into recipient thymus after bone marrow transplantation plus bone grafts: A role of stromal cells in positive selection. Exp Hematol (2000). , 28(2000), 950-60.

[24] Saito, T, Kuang, J. Q, Lin, C. C, & Chiu, R. C. Transcoronary implantation of bone marrow stromal cells ameliorates cardiac function after myocardial infarction. J Thorac Cardiovasc Surg (2003). , 126(2003), 114-23.

[25] Phinney, D. G, Baddoo, M, Dutreil, M, Gaupp, D, Lai, W. T, & Isakova, I. A. Murine mesenchymal stem cells transplanted to the central nervous system of neonatal versus adult mice exhibit distinct engraftment kinetics and express receptors that guide neuronal cell migration. Stem Cells Dev (2006). , 15(2006), 437-47.

[26] Li, Y, Hisha, H, Inaba, M, Lian, Z, Yu, C, Kawamura, M, Yamamoto, Y, Nishio, N, Toki, J, Fan, H, & Ikehara, S. Evidence for migration of donor bone marrow stromal cells into recipient thymus after bone marrow transplantation plus bone grafts: A role of stromal cells in positive selection. Exp Hematol (2000). , 28(2000), 950-60.

[27] Liechty, K.W, & Mac, T.C. . Marshak, and A.W. Flake, Human mesenchymal stem cells engraft and demonstrate site-specific differentiation after in utero transplantation in sheep. Nat Med 6 (2000) 1282-6.

[28] Airey, J. A, Almeida-porada, G, Colletti, E. J, Porada, C. D, Chamberlain, J, Movsesian, M, Sutko, J. L, & Zanjani, E. D. Human mesenchymal stem cells form Purkinje fibers in fetal sheep heart. Circulation (2004). , 109(2004), 1401-7.

[29] Herrera, M. B, Bussolati, B, Bruno, S, Fonsato, V, Romanazzi, G. M, & Camussi, G. Mesenchymal stem cells contribute to the renal repair of acute tubular epithelial injury. Int J Mol Med (2004). , 14(2004), 1035-41.

[30] Ortiz, L. A, Gambelli, F, Mcbride, C, Gaupp, D, Baddoo, M, Kaminski, N, & Phinney, D. G. Mesenchymal stem cell engraftment in lung is enhanced in response to bleomycin exposure and ameliorates its fibrotic effects. Proc Natl Acad Sci U S A (2003). , 100(2003), 8407-11.

[31] Mangi, A. A, Noiseux, N, Kong, D, He, H, Rezvani, M, Ingwall, J. S, & Dzau, V. J. Mesenchymal stem cells modified with Akt prevent remodeling and restore performance of infarcted hearts. Nat Med (2003). , 9(2003), 1195-201.

[32] Mahmud, N, Pang, W, Cobbs, C, Alur, P, Borneman, J, Dodds, R, Archambault, M, Devine, S, Turian, J, Bartholomew, A, Vanguri, P, Mackay, A, Young, R, & Hoffman, R. Studies of the route of administration and role of conditioning with radiation on unrelated allogeneic mismatched mesenchymal stem cell engraftment in a nonhuman primate model. Exp Hematol (2004). , 32(2004), 494-501.

[33] Lazarus, H. M, Haynesworth, S. E, Gerson, S. L, Rosenthal, N. S, & Caplan, A. I. Ex vivo expansion and subsequent infusion of human bone marrow-derived stromal progenitor cells (mesenchymal progenitor cells): implications for therapeutic use. Bone Marrow Transplant (1995). , 16(1995), 557-64.

[34] Wang, J, Liu, K, & Lu, D. P. Mesenchymal stem cells in stem cell transplant recipients are damaged and remain of host origin. Int J Hematol (2005). , 82(2005), 152-8.

[35] Le, K, Blanc, C, Gotherstrom, O, Ringden, M, Hassan, R, Mcmahon, E, Horwitz, G, Anneren, O, Axelsson, J, Nunn, U, Ewald, S, Norden-lindeberg, M, Jansson, A, & Dalton, E. Astrom, and M. Westgren, Fetal mesenchymal stem-cell engraftment in bone after in utero transplantation in a patient with severe osteogenesis imperfecta. Transplantation (2005). , 79(2005), 1607-14.

[36] Fouillard, L, Bensidhoum, M, Bories, D, Bonte, H, Lopez, M, Moseley, A. M, Smith, A, Lesage, S, Beaujean, F, Thierry, D, Gourmelon, P, Najman, A, & Gorin, N. C. Engraftment of allogeneic mesenchymal stem cells in the bone marrow of a patient with severe idiopathic aplastic anemia improves stroma. Leukemia (2003). , 17(2003), 474-6.

[37] Dickhut, A, Schwerdtfeger, R, Kuklick, L, Ritter, M, Thiede, C, Neubauer, A, & Brendel, C. Mesenchymal stem cells obtained after bone marrow transplantation or peripheral blood stem cell transplantation originate from host tissue. Ann Hematol (2005). , 84(2005), 722-7.

[38] Rubio, D, Garcia, S, Paz, M. F, De La Cueva, T, Lopez-fernandez, L. A, Lloyd, A. C, Garcia-castro, J, & Bernad, A. Molecular characterization of spontaneous mesenchymal stem cell transformation. PLoS ONE 3 ((2008). e1398.

[39] Rubio, D, Garcia-castro, J, Martin, M. C, De La Fuente, R, Cigudosa, J. C, Lloyd, A. C, & Bernad, A. Spontaneous human adult stem cell transformation. Cancer Res (2005). , 65(2005), 3035-9.

[40] Tolar, J, Nauta, A. J, Osborn, M. J, Panoskaltsis, A, Mortari, R. T, Mcelmurry, S, Bell, L, Xia, N, Zhou, M, Riddle, T. M, Schroeder, J. J, Westendorf, R. S, Mcivor, P. C, Hogendoorn, K, Szuhai, L, Oseth, B, Hirsch, S. R, Yant, M. A, Kay, A, Peister, D. J, & Prockop, W. E. Fibbe, and B.R. Blazar, Sarcoma derived from cultured mesenchymal stem cells. Stem Cells (2007). , 25(2007), 371-9.

[41] Bernardo, M. E, Zaffaroni, N, Novara, F, Cometa, A. M, Avanzini, M. A, Moretta, A, Montagna, D, Maccario, R, Villa, R, Daidone, M. G, Zuffardi, O, & Locatelli, F. Human bone marrow derived mesenchymal stem cells do not undergo transformation after long-term in vitro culture and do not exhibit telomere maintenance mechanisms. Cancer Res (2007). , 67(2007), 9142-9.

[42] Houghton, J, Stoicov, C, Nomura, S, Rogers, A. B, Carlson, J, Li, H, Cai, X, Fox, J. G, Goldenring, J. R, & Wang, T. C. Gastric cancer originating from bone marrow-derived cells. Science (2004). , 306(2004), 1568-71.

[43] Miura, M, Miura, Y, Padilla-nash, H. M, Molinolo, A. A, Fu, B, Patel, V, Seo, B. M, Sonoyama, W, Zheng, J. J, Baker, C. C, Chen, W, Ried, T, & Shi, S. Accumulated chromosomal instability in murine bone marrow mesenchymal stem cells leads to malignant transformation. Stem Cells (2006). , 24(2006), 1095-103.

[44] Rebecca, S. Y. Wong. Review Articl. Mesenchymal StemCells: Angels or Demons? Journal of Biomedicine and Biotechnology. Volume, Article ID 459510, 8 pages

[45] Karnoub, A. E, Dash, A. B, Vo, A. P, Sullivan, A, Brooks, M. W, Bell, G. W, Richardson, A. L, Polyak, K, Tubo, R, & Weinberg, R. A. Mesenchymal stem cells within tumour stroma promote breast cancer metastasis. Nature (2007). , 449(2007), 557-63.

[46] Lazennec, G, & Jorgensen, C. Adult Multipotent Stromal Cells and Cancer: Risk or Benefit? Stem Cells ((2008).

[47] Noort, W. A, Kruisselbrink, A. B, In, P. S, Anker, t, Kruger, M, Van Bezooijen, R. L, De Paus, R. A, Heemskerk, M. H, Lowik, C. W, Falkenburg, J. H, Willemze, R, & Fibbe, W. E. Mesenchymal stem cells promote engraftment of human umbilical cord blood-derived CD34(+) cells in NOD/SCID mice. Exp Hematol (2002). , 30(2002), 870-8.

[48] Anklesaria, P, Kase, K, Glowacki, J, Holland, C.A, Sakakeeny, M.A, Wright, J.A, & Fitz, T.J. . Lee, and J.S. Greenberger, Engraftment of a clonal bone marrow stromal cell line in vivo stimulates hematopoietic recovery from total body irradiation. Proc Natl Acad Sci U S A 84 (1987) 7681-5.

[49] Bartholomew, A, Sturgeon, C, Siatskas, M, Ferrer, K, Mcintosh, K, Patil, S, Hardy, W, Devine, S, Ucker, D, Deans, R, Moseley, A, & Hoffman, R. Mesenchymal stem cells suppress lymphocyte proliferation in vitro and prolong skin graft survival in vivo. Exp Hematol (2002). , 30(2002), 42-8.

[50] Le, K, Blanc, I, Rasmusson, B, Sundberg, C, Gotherstrom, M, & Hassan, M. Uzunel, and O. Ringden, Treatment of severe acute graft-versus-host disease with third party haploidentical mesenchymal stem cells. Lancet (2004). , 363(2004), 1439-41.

[51] Ildstad, S. T, & Sachs, D. H. Reconstitution with syngeneic plus allogeneic or xenogeneic bone marrow leads to specific acceptance of allografts or xenografts. Nature (1984). , 307(1984), 168-70.

[52] Gurevitch, O, Prigozhina, T. B, Pugatsch, T, & Slavin, S. Transplantation of allogeneic or xenogeneic bone marrow within the donor stromal microenvironment. Transplantation (1999). , 68(1999), 1362-8.

[53] Bingaman, A. W, Waitze, S. Y, Alexander, D. Z, Cho, H. R, Lin, A, Tucker-burden, C, Cowan, S. R, Pearson, T. C, & Larsen, C. P. Transplantation of the bone marrow microenvironment leads to hematopoietic chimerism without cytoreductive conditioning. Transplantation (2000). , 69(2000), 2491-6.

[54] Sakaguchi, Y, Sekiya, I, Yagishita, K, Ichinose, S, Shinomiya, K, & Muneta, T. Suspended cells from trabecular bone by collagenase digestion become virtually identical to mesenchymal stem cells obtained from marrow aspirates. Blood (2004). , 104(2004), 2728-35.

[55] Yanez, R, Lamana, M. L, Garcia-castro, J, Colmenero, I, Ramirez, M, & Bueren, J. A. Adipose tissue-derived mesenchymal stem cells have in vivo immunosuppressive properties applicable for the control of the graft-versus-host disease. Stem Cells (2006). , 24(2006), 2582-91.

[56] Sudres, M, Norol, F, Trenado, A, Gregoire, S, Charlotte, F, Levacher, B, Lataillade, J. J, Bourin, P, Holy, X, Vernant, J. P, Klatzmann, D, & Cohen, J. L. Bone marrow mesenchymal stem cells suppress lymphocyte proliferation in vitro but fail to prevent graft-versus-host disease in mice. J Immunol (2006). , 176(2006), 7761-7.

[57] Djouad, F, Fritz, V, Apparailly, F, Louis-plence, P, Bony, C, Sany, J, Jorgensen, C, & Noel, D. Reversal of the immunosuppressive properties of mesenchymal stem cells by tumor necrosis factor alpha in collagen-induced arthritis. Arthritis Rheum (2005). , 52(2005), 1595-603.

[58] Maitra, B, Szekely, E, Gjini, K, Laughlin, M. J, Dennis, J, Haynesworth, S. E, & Koc, O. N. Human mesenchymal stem cells support unrelated donor hematopoietic stem cells and suppress T-cell activation. Bone Marrow Transplant (2004). , 33(2004), 597-604.

[59] Chung, N. G, Jeong, D. C, Park, S. J, Choi, B. O, Cho, B, Kim, H. K, Chun, C. S, Won, J. H, & Han, C. W. Cotransplantation of marrow stromal cells may prevent lethal graft-versus-host disease in major histocompatibility complex mismatched murine hematopoietic stem cell transplantation. Int J Hematol (2004). , 80(2004), 370-6.

[60] Xu, G, Zhang, L, Ren, G, Yuan, Z, Zhang, Y, Zhao, R. C, & Shi, Y. Immunosuppressive properties of cloned bone marrow mesenchymal stem cells. Cell Res (2007). , 17(2007), 240-8.

[61] Eliopoulos, N, Stagg, J, Lejeune, L, Pommey, S, & Galipeau, J. Allogeneic marrow stromal cells are immune rejected by MHC class I- and class II-mismatched recipient mice. Blood (2005). , 106(2005), 4057-65.

[62] Verfaillie, C. M. Adhesion receptors as regulators of the hematopoietic process. Blood (1998)., 92(1998), 2609-12.

[63] Arroyo, A. G, Yang, J. T, Rayburn, H, & Hynes, R. O. Alpha4 integrins regulate the proliferation/differentiation balance of multilineage hematopoietic progenitors in vivo. Immunity (1999)., 11(1999), 555-66.

[64] Koc, O. N, Gerson, S. L, Cooper, B. W, Dyhouse, S. M, Haynesworth, S. E, Caplan, A. I, & Lazarus, H. M. Rapid hematopoietic recovery after coinfusion of autologous-blood stem cells and culture-expanded marrow mesenchymal stem cells in advanced breast cancer patients receiving high-dose chemotherapy. J Clin Oncol (2000). , 18(2000), 307-16.

[65] Lee, S. T, Jang, J. H, Cheong, J. W, Kim, J. S, Maemg, H. Y, Hahn, J. S, Ko, Y. W, & Min, Y. H. Treatment of high-risk acute myelogenous leukaemia by myeloablative chemoradiotherapy followed by co-infusion of T cell-depleted haematopoietic stem cells and culture-expanded marrow mesenchymal stem cells from a related donor with one fully mismatched human leucocyte antigen haplotype. Br J Haematol (2002)., 118(2002), 1128-31.

[66] Kim, D. W, Chung, Y. J, Kim, T. G, Kim, Y. L, & Oh, I. H. Cotransplantation of third-party mesenchymal stromal cells can alleviate single-donor predominance and increase engraftment from double cord transplantation. Blood (2004). , 103(2004), 1941-8.

[67] Ball, L. M, Bernardo, M. E, Roelofs, H, Lankester, A, Cometa, A, Egeler, M, Giorgiani, G, Locatelli, F, & Co-transplantation, F. W. E. of HLA-Haploidentical, Bone Marrow Derived Mesenchymal Stem Cells Prevents Graft Failure and Improves Hematological Recovery in T-Cell Depleted Haploidentical Stem Cell Transplantation. Blood 108 ((2006).

[68] Fouillard, L, Chapel, A, Bories, D, Bouchet, S, Costa, J. M, Rouard, H, Herve, P, Gourmelon, P, Thierry, D, Lopez, M, & Gorin, N. C. Infusion of allogeneic-related HLA mismatched mesenchymal stem cells for the treatment of incomplete engraftment following autologous haematopoietic stem cell transplantation. Leukemia ((2007).

[69] Le, K, Blanc, H, Samuelsson, B, Gustafsson, M, Remberger, B, Sundberg, J, Arvidson, P, Ljungman, H, & Lonnies, S. Nava, and O. Ringden, Transplantation of mesenchymal stem cells to enhance engraftment of hematopoietic stem cells. Leukemia (2007)., 21(2007), 1733-8.

[70] Sundin, M, Ringden, O, Sundberg, B, Nava, S, & Gotherstrom, C. and K. Le Blanc, No alloantibodies against mesenchymal stromal cells, but presence of anti-fetal calf serum antibodies, after transplantation in allogeneic hematopoietic stem cell recipients. Haematologica (2007)., 92(2007), 1208-15.

[71] Aggarwal, S, & Pittenger, M. F. Human mesenchymal stem cells modulate allogeneic immune cell responses. Blood (2005)., 105(2005), 1815-22.

[72] Frassoni, l. M, & Bacigalupo, F. A, et al., Expanded mesenchymql stem cells (MSC) co-infused with HLA-identical hematopoietic stem cells transplants, reduce acute and chronic graft-vs-host disease: a matched pair analysis.. Bone Marrow Transplant 29 ((2002).

[73] Le Blanc KFrassoni F, Ball LM, Lanino E, Sundberg B, Lonnies L, Roelofs H, Dini G, Bacigalupo A, Locatelli F, Fibbe WF, and O. Ringden, Mesenchymal Stem Cells for Treatment of Severe Acute Graft-Versus-Host Disease.. Blood 108 ((2006).

[74] Lazarus, H. M, Koc, O. N, Devine, S. M, Curtin, P, Maziarz, R. T, Holland, H. K, Shpall, E. J, Mccarthy, P, Atkinson, K, Cooper, B. W, Gerson, S. L, Laughlin, M. J, Loberiza, F. R, Jr, A. B, & Moseley, A. Bacigalupo, Cotransplantation of HLA-identical sibling culture-expanded mesenchymal stem cells and hematopoietic stem cells in hematologic malignancy patients. Biol Blood Marrow Transplant (2005). , 11(2005), 389-98.

[75] Ringden, O, Uzunel, M, Rasmusson, I, Remberger, M, Sundberg, B, Lonnies, H, Marschall, H. U, Dlugosz, A, Szakos, A, Hassan, Z, Omazic, B, Aschan, J, & Barkholt, L. and K. Le Blanc, Mesenchymal stem cells for treatment of therapy-resistant graft-versus-host disease. Transplantation (2006). , 81(2006), 1390-7.

[76] Ringden, O, Uzunel, M, Sundberg, B, Lonnies, L, Nava, S, Gustafsson, J, & Henningsohn, L. and K. Le Blanc, Tissue repair using allogeneic mesenchymal stem cells for hemorrhagic cystitis, pneumomediastinum and perforated colon. Leukemia (2007). , 21(2007), 2271-6.

[77] Pérez-simon, J. A, López-villar, O, Andreu, E. J, Rifón, J, Muntion, S, Campelo, M. D, Sánchez-guijo, F. M, Martinez, C, & Valcarcel, D. and del zo C. Mesenchymal stem cellsexpanded in vitro with human serum for the treatment of acuteand chronic graft-versus-host disease: results phase I/II clinical trial. Haematologica (2011). , 96(07), 1072-1076.

[78] Lazarus, H. M, Haynesworth, S. E, Gerson, S. L, Rosenthal, N. S, & Caplan, A. I. Ex vivo expansion and subsequent infusion of human bone marrow-derived stromal progenitor cells (mesenchymal progenitor cells): implications for therapeutic use. Bone Marrow Transplant (1995). , 16(1995), 557-64.

[79] Djouad, F, Plence, P, Bony, C, Tropel, P, Apparailly, F, Sany, J, Noel, D, & Jorgensen, C. Immunosuppressive effect of mesenchymal stem cells favors tumor growth in allogeneic animals. Blood (2003). , 102(2003), 3837-44.

[80] Nasef, A, Chapel, A, Mazurier, C, Bouchet, S, Lopez, M, Mathieu, N, Sensebe, L, Zhang, Y, Gorin, N. C, Thierry, D, & Fouillard, L. Identification of IL-10 and TGF-beta transcripts involved in the inhibition of T-lymphocyte proliferation during cell contact with human mesenchymal stem cells. Gene Expr (2007). , 13(2007), 217-26.

[81] Di, M, Nicola, C, Carlo-stella, M, Magni, M, Milanesi, P. D, Longoni, P, & Matteucci, S. Grisanti, and A.M. Gianni, Human bone marrow stromal cells suppress T-lympho-

cyte proliferation induced by cellular or nonspecific mitogenic stimuli. Blood (2002)., 99(2002), 3838-43.

[82] Caplan, A. I, & Dennis, J. E. Mesenchymal stem cells as trophic mediators. J Cell Biochem (2006)., 98(2006), 1076-84.

[83] Kim, D. H, Yoo, K. H, Choi, K. S, Choi, J, Choi, S. Y, Yang, S. E, Yang, Y. S, Im, H. J, Kim, K. H, Jung, H. L, Sung, K. W, & Koo, H. H. Gene expression profile of cytokine and growth factor during differentiation of bone marrow-derived mesenchymal stem cell. Cytokine (2005)., 31(2005), 119-26.

[84] Steckel, N. K, Kuhn, U, Beelen, D. W, & Elmaagacli, A. H. Indoleamine 2,3-dioxygenase expression in patients with acute graft-versus-host disease after allogeneic stem cell transplantation and in pregnant women: association with the induction of allogeneic immune tolerance? Scand J Immunol (2003)., 57(2003), 185-91.

[85] Le, K, Blanc, L, Tammik, B, & Sundberg, S. E. Haynesworth, and O. Ringden, Mesenchymal stem cells inhibit and stimulate mixed lymphocyte cultures and mitogenic responses independently of the major histocompatibility complex. Scand J Immunol (2003)., 57(2003), 11-20.

[86] Plumas, J, Chaperot, L, Richard, M. J, Molens, J. P, Bensa, J. C, & Favrot, M. C. Mesenchymal stem cells induce apoptosis of activated T cells. Leukemia (2005)., 19(2005), 1597-604.

[87] Glennie, S, Soeiro, I, Dyson, P. J, Lam, E. W, & Dazzi, F. Bone marrow mesenchymal stem cells induce division arrest anergy of activated T cells. Blood (2005)., 105(2005), 2821-7.

[88] Maccario, R, Podesta, M, Moretta, A, Cometa, A, Comoli, P, Montagna, D, Daudt, L, Ibatici, A, Piaggio, G, Pozzi, S, Frassoni, F, & Locatelli, F. Interaction of human mesenchymal stem cells with cells involved in alloantigen-specific immune response favors the differentiation of CD4+ T-cell subsets expressing a regulatory/suppressive phenotype. Haematologica (2005)., 90(2005), 516-25.

[89] Krampera, M, Glennie, S, Dyson, J, Scott, D, Laylor, R, Simpson, E, & Dazzi, F. Bone marrow mesenchymal stem cells inhibit the response of naive and memory antigen-specific T cells to their cognate peptide. Blood (2003)., 101(2003), 3722-9.

[90] Angoulvant, D, Clerc, A, Benchalal, S, Galambrun, C, Farre, A, Bertrand, Y, & Eljaafari, A. Human mesenchymal stem cells suppress induction of cytotoxic response to alloantigens. Biorheology (2004)., 41(2004), 469-76.

[91] Sotiropoulou, P. A, Perez, S. A, Gritzapis, A. D, Baxevanis, C. N, & Papamichail, M. Interactions between human mesenchymal stem cells and natural killer cells. Stem Cells (2006)., 24(2006), 74-85.

[92] Miller, R. G, Muraoka, S, Claesson, M. H, Reimann, J, & Benveniste, P. The veto phenomenon in T-cell regulation. Ann N Y Acad Sci (1988)., 532(1988), 170-6.

[93] Potian, J. A, Aviv, H, Ponzio, N. M, Harrison, J. S, & Rameshwar, P. Veto-like activity of mesenchymal stem cells: functional discrimination between cellular responses to alloantigens and recall antigens. J Immunol (2003). , 171(2003), 3426-34.

[94] Rammensee, H. G, Fink, P. J, & Bevan, M. J. Functional clonal deletion of class I-specific cytotoxic T lymphocytes by veto cells that express antigen. J Immunol (1984). , 133(1984), 2390-6.

[95] Batten, P, Sarathchandra, P, Antoniw, J. W, Tay, S. S, Lowdell, M. W, Taylor, P. M, & Yacoub, M. H. Human mesenchymal stem cells induce T cell anergy and downregulate T cell allo-responses via the TH2 pathway: relevance to tissue engineering human heart valves. Tissue Eng (2006). , 12(2006), 2263-73.

[96] Chen, L, Zhang, W, Yue, H, Han, Q, Chen, B, Shi, M, Li, J, Li, B, You, S, Shi, Y, & Zhao, R. C. Effects of human mesenchymal stem cells on the differentiation of dendritic cells from CD34+ cells. Stem Cells Dev (2007). , 16(2007), 719-31.

[97] Raffaghello, L, Bianchi, G, Bertolotto, M, Montecucco, F, Busca, A, Dallegri, F, Ottonello, L, & Pistoia, V. Human mesenchymal stem cells inhibit neutrophil apoptosis: a model for neutrophil preservation in the bone marrow niche. Stem Cells (2008). , 26(2008), 151-62.

[98] Tse, W. T, Pendleton, J. D, Beyer, W. M, Egalka, M. C, & Guinan, E. C. Suppression of allogeneic T-cell proliferation by human marrow stromal cells: implications in transplantation. Transplantation (2003). , 75(2003), 389-97.

[99] Meisel, R, Zibert, A, Laryea, M, Gobel, U, Daubener, W, & Dilloo, D. Human bone marrow stromal cells inhibit allogeneic T-cell responses by indoleamine 2,3-dioxygenase-mediated tryptophan degradation. Blood (2004). , 103(2004), 4619-21.

[100] Moviglia, G. A, Varela, G, Gaeta, C. A, Brizuela, J. A, Bastos, F, & Saslavsky, J. Autoreactive T cells induce in vitro BM mesenchymal stem cell transdifferentiation to neural stem cells. Cytotherapy (2006). , 8(2006), 196-201.

[101] Groh, M. E, Maitra, B, Szekely, E, & Koc, O. N. Human mesenchymal stem cells require monocyte-mediated activation to suppress alloreactive T cells. Exp Hematol (2005). , 33(2005), 928-34.

[102] Rasmusson, I, Ringden, O, & Sundberg, B. and K. Le Blanc, Mesenchymal stem cells inhibit lymphocyte proliferation by mitogens and alloantigens by different mechanisms. Exp Cell Res (2005). , 305(2005), 33-41.

[103] Poggi, A, Prevosto, C, Massaro, A. M, Negrini, S, Urbani, S, Pierri, I, Saccardi, R, Gobbi, M, & Zocchi, M. R. Interaction between human NK cells and bone marrow stromal cells induces NK cell triggering: role of NKp30 and NKG2D receptors. J Immunol (2005). , 175(2005), 6352-60.

[104] Spaggiari, G. M, Capobianco, A, Abdelrazik, H, Becchetti, F, Mingari, M. C, & Moretta, L. Mesenchymal stem cells inhibit natural killer-cell proliferation, cytotoxicity,

and cytokine production: role of indoleamine 2,3-dioxygenase and prostaglandin E2. Blood (2008). , 111(2008), 1327-33.

[105] Ren, G, Zhang, L, Zhao, X, Xu, G, Zhang, Y, Roberts, A. I, Zhao, R. C, & Shi, Y. Mesenchymal Stem Cell-Mediated Immunosuppression Occurs via Concerted Action of Chemokines and Nitric Oxide. Cell Stem Cell (2008). , 2(2008), 141-150.

[106] Ning, H, Yang, F, Jiang, M, Hu, L, Feng, K, Zhang, J, Yu, z, Li, B, Xu, C, Li, Y, Wang, J, Hu, J, Lou, X, & Chen, H. The correlation between cotransplantation of mesenchymal stem cells and higher recurrence rate in hematologic malignancy patients: outcome of apilot clinical study. Leukemia 22 (3) ((2008).